Optical Communications and Networking

Optical Communications and Networking

Prospects in Industrial Applications

Special Issue Editors

Zhongqi Pan
Qiang Wang
Yang Yue
Hao Huang
Changjing Bao

MDPI • Basel • Beijing • Wuhan • Barcelona • Belgrade

Special Issue Editors

Zhongqi Pan
University of Louisiana at Lafayette
USA

Qiang Wang
Nokia Corporation
USA

Yang Yue
Nankai University
China

Hao Huang
Lumentum Operations LLC
USA

Changjing Bao
Nokia Corporation
USA

Editorial Office
MDPI
St. Alban-Anlage 66
4052 Basel, Switzerland

This is a reprint of articles from the Special Issue published online in the open access journal *Applied Sciences* (ISSN 2076-3417) from 2018 to 2020 (available at: https://www.mdpi.com/journal/applsci/special_issues/Optical_Communications_Networking_Industrial).

For citation purposes, cite each article independently as indicated on the article page online and as indicated below:

LastName, A.A.; LastName, B.B.; LastName, C.C. Article Title. *Journal Name* **Year**, *Article Number*, Page Range.

ISBN 978-3-03928-258-6 (Pbk)
ISBN 978-3-03928-259-3 (PDF)

© 2020 by the authors. Articles in this book are Open Access and distributed under the Creative Commons Attribution (CC BY) license, which allows users to download, copy and build upon published articles, as long as the author and publisher are properly credited, which ensures maximum dissemination and a wider impact of our publications.

The book as a whole is distributed by MDPI under the terms and conditions of the Creative Commons license CC BY-NC-ND.

Contents

About the Special Issue Editors vii

Zhongqi Pan, Qiang Wang, Yang Yue, Hao Huang and Changjing Bao
Special Issue on Optical Communications and Networking: Prospects in Industrial Applications
Reprinted from: *Appl. Sci.* **2020**, *10*, 411, doi:10.3390/app10010411 1

Tianhua Xu, Cenqin Jin, Shuqing Zhang, Gunnar Jacobsen, Sergei Popov, Mark Leeson and Tiegen Liu
Phase Noise Cancellation in Coherent Communication Systems Using a Radio Frequency Pilot Tone
Reprinted from: *Appl. Sci.* **2019**, *9*, 4717, doi:10.3390/app9214717 6

Yang Yue, Qiang Wang, Jian Yao, Jason O'Neil, Daniel Pudvay and Jon Anderson
400GbE Technology Demonstration Using CFP8 Pluggable Modules
Reprinted from: *Appl. Sci.* **2018**, *8*, 2055, doi:10.3390/app8112055 15

Jun Yeong Jang, Min Su Kim, Chang-Lin Li and Tae Hee Han
Power and Signal-to-Noise Ratio Optimization in Mesh-Based Hybrid Optical Network-on-Chip Using Semiconductor Optical Amplifiers
Reprinted from: *Appl. Sci.* **2019**, *9*, 1251, doi:10.3390/app9061251 24

Zhen Qu, Ivan B. Djordjevic and Jon Anderson
Two-Dimensional Constellation Shaping in Fiber-Optic Communications
Reprinted from: *Appl. Sci.* **2019**, *9*, 1889, doi:10.3390/app9091889 47

Xiao Han, Mingwei Yang, Ivan B. Djordjevic, Yang Yue, Qiang Wang, Zhen Qu and Jon Anderson
Joint Probabilistic-Nyquist Pulse Shaping for an LDPC-Coded 8-PAM Signal in DWDM Data Center Communications
Reprinted from: *Appl. Sci.* **2019**, *9*, 4996, doi:10.3390/app9234996 60

Qiang Wang, Yang Yue, Jian Yao and Jon Anderson
Adaptive Compensation of Bandwidth Narrowing Effect for Coherent In-Phase Quadrature Transponder through Finite Impulse Response Filter
Reprinted from: *Appl. Sci.* **2019**, *9*, 1950, doi:10.3390/app9091950 69

Yu Zuo and Jian Zhang
A Novel Coding Based Dimming Scheme with Constant Transmission Efficiency in VLC Systems
Reprinted from: *Appl. Sci.* **2019**, *9*, 803, doi:10.3390/app9040803 82

Simeng Feng, Hailiang Feng, Ying Zhou and Baolong Li
Low-Complexity Hybrid Optical OFDM with High Spectrum Efficiency forDimming Compatible VLC System
Reprinted from: *Appl. Sci.* **2019**, *9*, 3666, doi:10.3390/app9183666 89

Dong He, Qiang Wang, Xiang Liu, Zhijun Song, Jianwei Zhou, Zhongke Wang, Chunyang Gao, Tong Zhang, Xiaoping Qi, Yi Tan, Ge Ren, Bo Qi, Jigang Ren, Yuan Cao and Yongmei Huang
Shipborne Acquisition, Tracking, and Pointing Experimental Verifications towards Satellite-to-Sea Laser Communication
Reprinted from: *Appl. Sci.* **2019**, *9*, 3940, doi:10.3390/app9183940 . . . . **105**

About the Special Issue Editors

Zhongqi Pan received his B.S. and M.S. degrees from Tsinghua University, China, and his Ph.D. degree from the University of Southern California, Los Angeles, all in electrical engineering. He is currently a Professor at the Department of Electrical and Computer Engineering. He also holds BORSF Endowed Professorship in Electrical Engineering II, and BellSouth/BoRSF Endowed Professorship in Telecommunications. Dr. Pan's research is in the area of photonics, including photonic devices, fiber communications, wavelength-division-multiplexing (WDM) technologies, optical performance monitoring, coherent optical communications, space-division-multiplexing (SDM) technologies, and fiber-sensor technologies. He has authored/coauthored 160 publications, including five book chapters and >20 invited presentations/papers. He also has five U. S. patents and one China patent. Prof. Pan is an OSA and IEEE senior member.

Qiang Wang received his B.S. degree in electrical engineering from Huazhong University of Science and Technology, Wuhan, China, in 1995. He received his M.S. degree in electrical engineering from Chinese Academy of Science, Shanghai, China in 1998. He received his M.S. and Ph.D. degrees in electrical engineering from University of Maryland Baltimore County, Baltimore, Maryland, USA in 2000 and 2010, respectively. From 2000 to 2002, he was a Member of Technical Staff with Lucent Technologies. From 2007 to 2013, he was a Staff Hardware Engineer with Infinera Corporation. From 2013 to 2018, he was a Staff Hardware Engineer with Juniper Networks, Sunnyvale, CA, USA. Currently he is an Optical System Architect in Nokia Corporation, Mountain View, CA, USA. He is the author of 12 journal articles and 30 conference articles. He holds 10 US patents and has 20 pending patent applications. His research interests include optical communications and networking, integrated photonics, fiber optics, free-space optics, and machine learning. Dr. Wang is a Member of the Optical Society of America (OSA) and the International Society for Optical Engineering (SPIE).

Yang Yue received his B.S. and M.S. degrees in electrical engineering and optics from Nankai University, Tianjin, China, in 2004 and 2007, respectively. He received his Ph.D. degree in electrical engineering from the University of Southern California, Los Angeles, CA, USA, in 2012. He is a Professor with the Institute of Modern Optics, Nankai University, Tianjin, China. Dr. Yue's current research interests include intelligent photonics, optical communications and networking, optical interconnect, detection, imaging, and display technology. He has published over 150 peer-reviewed journal papers and conference proceedings, three edited books, one book chapter, >10 invited papers, >30 issued or pending patents, and >80 invited presentations.

Hao Huang received his B.S. degree from Jilin University, Changchun, China, in 2006, and his M.S. degree from Beijing University of Posts and Telecommunications, Beijing, China, in 2009. He received his Ph.D. degree in electrical engineering from University of Southern California, Los Angeles, California, USA, in 2014. He is currently working at Lumentum Operations LLC as a hardware engineer. His research areas include optical communication system and components, optical sensing systems, and digital signal processing. He has coauthored more than 100 publications, including peer-reviewed journals and conference proceedings. He is a member of the Optical Society of America (OSA).

Changjing Bao received his Ph.D. degree in electrical engineering from University of Southern California, Los Angeles, California, USA, in 2017. He is currently working at Nokia, New Jersey, as an optical engineer. His research interests include optical communications, nonlinear optics, and integrated optics. He has authored and coauthored more than 100 journal papers and conference proceedings.

Editorial

Special Issue on Optical Communications and Networking: Prospects in Industrial Applications

Zhongqi Pan [1], Qiang Wang [2], Yang Yue [3,*], Hao Huang [4] and Changjing Bao [5]

1. Department of Electrical and Computer Engineering, University of Louisiana at Lafayette, Lafayette, LA 70504, USA; zhongqi.pan@louisiana.edu
2. Nokia Corporation, Mountain View, CA 94043, USA; qiwang.thresh@gmail.com
3. Institute of Modern Optics, Nankai University, Tianjin 300350, China
4. Lumentum Operations LLC, San Jose, CA 95131, USA; hi.haoh@gmail.com
5. Nokia Corporation, Murray Hill, NJ 07974, USA; baochangjing@gmail.com
* Correspondence: yueyang@nankai.edu.cn

Received: 18 December 2019; Accepted: 25 December 2019; Published: 6 January 2020

1. Introduction

In the past two decades, Internet traffic has increased by over 10,000 times by taking advantage of both efficient information processing technology in the electronic domain and efficient transmission technology in the optical domain, which are the foundation of today's Internet infrastructure [1,2]. The advancement of electronics processing circuits has followed Moore's law, and perhaps will continue this exponential growth for years to come. This may make the electrical systems significantly outpace the advancement of optical systems in information and communications technologies. To support the ever-growing Internet traffic, optical communication systems face a great challenge in transporting information processed by electronic systems for sustained exponential growth. The industry has explored multiple degrees of freedom of the photon (time, wavelength, amplitude, phase, polarization, and space) to significantly reduce the cost/bit for data transmission by increasing the capacity/fiber through multiplexing and reducing the size and power through integration.

This Special Issue aims to explore the latest advancements in the optical communication industry. The applications range from short-reach chip-to-chip interconnections to the long-haul backbone communications at the trans-oceanic distance, including the technologies used in devices, systems, and networks levels. It focuses on state-of-the-art advances and future perspectives in commercially deployed systems.

2. Special Issue Papers

In this Special Issue, we have carefully selected nine articles, including one invited and eight contributed papers. These papers cover advanced techniques for short-reach applications, such as data centers, long-distance applications such as trans-oceanic links from systems that use single mode fibers, and the systems use free-space transmission, such as visible light communication (VLC) systems and satellite-to-sea laser communications.

The following are the summary of these papers:

- Phase Noise Cancellation in Coherent Communication Systems Using a Radio Frequency Pilot Tone [3]

In long-haul optical fiber communication, DSP-based (digital signal processing) dispersion compensation can be distorted by the equalization-enhanced phase noise (EEPN), due to the reciprocities between the dispersion compensation unit and the local oscillator (LO) laser phase noise (LPN). In this invited paper, the authors demonstrated a unique approach for phase noise cancellation using a radio frequency pilot tone. They showed that the RF pilot tone can entirely eliminate the LPN and efficiently suppress the EEPN when it is applied prior to the carrier phase recovery.

- 400GbE Technology Demonstration Using CFP8 Pluggable Modules [4]

In this paper, the authors reviewed the current status of 400GBASE client-side optics standards and multi-source agreements (MSAs). They then compared different form factors for 400GE modules, including CFP8, OSFP, and QSFP-DD. The essential techniques used to implement 400GE, such as pulse amplitude modulation (PAM4), forward error correction (FEC), and a continuous time-domain linear equalizer (CTLE), are also discussed. In addition, a 400GE physical interface card (PIC) in Juniper's PTX5000 platform has been developed, conforming to the latest IEEE802.3bs standard. To validate the PIC's performance, a commercial optical network tester (ONT) and the PIC are optically interconnected through two CFP8-LR8 modules. The CFP8-LR8 module utilizes eight optical wavelengths through coarse wavelength division multiplexing (CWDM). Each wavelength carries a 50 Gb/s PAM4 signal. The signal transmits through 10 km of single mode fiber (SMF). The ONT generates framed 400GE signal and sends it to the PIC through the first CFP8 module. The PIC recovers the signal, performs an internal loopback, and sends 400GE signal back to the ONT through the second CFP8 module. The optical spectrum, eye diagram, receiver sensitivity, long time soaking results, and internal digital diagnosis monitoring (DDM) result are fully characterized. The pre-FEC bit error rate (BER) is well below the KP4 FEC threshold of 2.2×10^{-4}. After KP4 FEC, an error-free performance over 30 km of SMF is achieved. At the end, the authors demonstrated both the interoperation between the PIC and the ONT, as well as the interoperation between the two CFP8 modules. This demonstration represents the successful implementation of the 400GE interface in the core IP/MPLS router.

- Power and Signal-to-Noise Ratio Optimization in Mesh-Based Hybrid Optical Network-on-Chip Using Semiconductor Optical Amplifiers [5]

To address the performance bottleneck in metal-based interconnects, hybrid optical network-on-chip (HONoC) has emerged as a new alternative. However, as the size of the HONoC grows, insertion loss and crosstalk noise increase, leading to excessive laser source output power and performance degradation. In this paper, the authors proposed a low-power scalable HONoC architecture by incorporating semiconductor optical amplifiers (SOAs). An SOA placement algorithm is developed considering insertion loss and crosstalk noise. Moreover, the authors establish a worst-case crosstalk noise model of SOA-enabled HONoC and induce optimized SOA gains with respect to power consumption and performance, respectively. Simulation results show that the proposed SOA-enabled HONoC architecture and the associated algorithm help sustain the performance as network size increases without the need for additional laser source power.

- Two-Dimensional Constellation Shaping in Fiber-Optic Communications [6]

Constellation shaping has been widely used in optical communication systems. The authors reviewed recent advances in two-dimensional constellation shaping technologies for fiber-optic communications. The system architectures that are discussed include probabilistic shaping, geometric shaping, and hybrid probabilistic-geometric shaping solutions. The performances of the three shaping schemes are also evaluated for Gaussian-noise-limited channels.

- Joint Probabilistic-Nyquist Pulse Shaping for an LDPC-Coded 8-PAM Signal in DWDM Data Center Communications [7]

M-ary pulse-amplitude modulation (PAM) has useful applications in data center communication due to its simplicity and low cost. The challenge in PAM systems include providing dynamic bandwidth and reaching the Shannon capacity limit. One potential solution is probabilistic shaping distribution with Nyquist pulse shaping for fine entropy granularity and reaching the Shannon limit. In this paper, the authors demonstrated the joint usage of probabilistic shaping and Nyquist pulse shaping with low-density parity-check (LDPC) coding to improve the bit error rate (BER) performance of 8-PAM signal transmission. They optimized the code rate of the LDPC code and investigated different Nyquist pulse shaping parameters using simulations and experiments. They achieved a 0.43 dB gain using Nyquist pulse shaping, and a 1.1 dB gain using probabilistic shaping, while the joint use of probabilistic shaping and Nyquist pulse shaping achieved a 1.27 dB gain, which offers an excellent improvement without upgrading the transceivers.

- Adaptive Compensation of Bandwidth Narrowing Effect for Coherent In-Phase Quadrature Transponder through Finite Impulse Response Filter [8]

In optical coherent communication systems, the transmitted signals may experience bandwidth narrowing effects after passing through multiple reconfigurable optical add-drop multiplexers (ROADMs), or due to coherent in-phase quadrature (IQ) transponder aging. In this paper, the authors demonstrated a method using a post ADC adaptive finite impulse response (FIR) filter in coherent optical receivers to dynamically compensate for the bandwidth narrowing effect. The influence of chromatic dispersion, polarization mode dispersion, and polarization dependent loss were also investigated comprehensively. Furthermore, the bandwidth information of the transmitted analog signal can be fed back to the coherent optical transmitter for signal optimization, and the transmitter-side FIR filter thus can be changed accordingly.

- A Novel Coding Based Dimming Scheme with Constant Transmission Efficiency in VLC Systems [9]

Visible light communications (VLC) has attracted tremendous attention due to two functions: communication and illumination. Both reliable data transmission and lighting quality need to be considered when the transmitted signal is designed. To achieve the desired levels of illumination, dimming control is an essential technology applied in VLC systems. In this paper, the authors proposed a block coding-based dimming scheme to construct the codeword set, where dimming control can be achieved by changing the ratio of two levels (ON and OFF) based on on-off keying (OOK) modulation. Simulation results show that the proposed scheme can maintain good error performance with constant transmission efficiency under various dimming levels.

- Low-Complexity Hybrid Optical OFDM with High Spectrum Efficiency for Dimming Compatible VLC System [10]

In this paper, the authors proposed a novel dimming compatible hybrid optical orthogonal frequency division multiplexing (DCHO-OFDM) method to fulfil the requirements for both communications and illuminations. Explicitly, the signal branch of the unclipped asymmetrically clipped O-OFDM (ACO-OFDM) and the down/upper-clipped pulse-amplitude-modulated discrete multitone (PAM-DMT) are adaptively combined to increase the spectrum efficiency. The chromaticity-shift-free and pulse width modulation (PWM) are adopted for the precisely dimming control, and a time-varying biasing scheme is used to mitigate the non-linear distortion. As the different signal components in DCHO-OFDM are combined in an interference-orthogonal approach, the transmitted symbols can be readily detected by a standard OFDM receiver. The simulation demonstrated that a high spectrum efficiency of the conceived DCHO-OFDM scheme can be achieved with less fluctuation in a wide dimming range.

- Shipborne Acquisition, Tracking, and Pointing Experimental Verifications towards Satellite-to-Sea Laser Communication [11]

Acquisition, tracking, and pointing (ATP) with high precision is a key technology in free space laser communication. In this paper, the authors report the acquisition and tracking of low-Earth-orbit satellites using shipborne ATP and verify the feasibility of establishing optical links between laser communication satellites and ships in the future. They developed a shipborne ATP system for satellite-to-sea applications in laser communications. The proposed acquisition strategy improves shipborne ATP pointing accuracy. They acquired and tracked some low-Earth-orbit satellites at sea, achieving a tracking accuracy of about 20 μrad. The results achieved in this work experimentally demonstrate the feasibility of ATP in satellite-to-sea laser communications.

3. Future Perspectives

Optics in the information and communication industry has a wide range of applications. It has experienced enormous growth in the past decades. Current communication systems and networks utilize the preeminent property of lightwaves for information transmission. To support the continuous exponential Internet traffic growth for many newly-emerging and unanticipated applications in the future, such as new services in the 5G network, we anticipate that the evolution of optical transmission technology will continue in order to meet the challenge ahead, probably through parallel process and integration [1,12]. On the other hand, many other emerging technologies, such as optical quantum communication and optical computing, promise to provide a number of new services and applications, and possibly to change future information infrastructure significantly [13]. This certainly requires more innovations and breakthroughs in optical signal processing technology.

Acknowledgments: First of all, the guest editors would like to thank all the authors for their excellent contributions to this special issue. Secondly, we would like to thank all the reviewers for their outstanding job in evaluating the manuscripts and providing valuable comments. Additionally, the guest editors would like to thank the MDPI team involved in the preparation, editing, and managing of this special issue. Finally, we would like to express our sincere gratitude to Lucia Li, the contact editor of this special issue, for her kind, efficient, professional guidance and support through the whole process. It would not be possible to have the above collection of high quality papers without these joint efforts.

Conflicts of Interest: The authors declare no conflict of interest.

References

1. Winzer, P.; Neilson, D.; Chraplyvy, A. Fiber-optic transmission and networking: The previous 20 and the next 20 years. *Opt. Express* **2018**, *26*, 24190–24239. [CrossRef] [PubMed]
2. Cheng, Q.; Bahadori, M.; Glick, M.; Rumley, S.; Bergman, K. Recent advances in optical technologies for data centers: A review. *Optica* **2018**, *5*, 1354–1370. [CrossRef]
3. Xu, T.; Jin, C.; Zhang, S.; Jacobsen, G.; Popov, S.; Leeson, M.; Liu, T. Phase Noise Cancellation in Coherent Communication Systems Using a Radio Frequency Pilot Tone. *Appl. Sci.* **2019**, *9*, 4717. [CrossRef]
4. Yue, Y.; Wang, Q.; Yao, J.; O'Neil, J.; Pudvay, D.; Anderson, J. 400GbE Technology Demonstration Using CFP8 Pluggable Modules. *Appl. Sci.* **2018**, *8*, 2055. [CrossRef]
5. Jang, J.Y.; Kim, M.S.; Li, C.-L.; Han, T.H. Power and Signal-to-Noise Ratio Optimization in Mesh-Based Hybrid Optical Network-on-Chip Using Semiconductor Optical Amplifiers. *Appl. Sci.* **2019**, *9*, 1251. [CrossRef]
6. Qu, Z.; Djordjevic, I.B.; Anderson, J. Two-Dimensional Constellation Shaping in Fiber-Optic Communications. *Appl. Sci.* **2019**, *9*, 1889. [CrossRef]
7. Han, X.; Yang, M.; Djordjevic, I.B.; Yue, Y.; Wang, Q.; Qu, Z.; Anderson, J. Joint Probabilistic-Nyquist Pulse Shaping for an LDPC-Coded 8-PAM Signal in DWDM Data Center Communications. *Appl. Sci.* **2019**, *9*, 4996. [CrossRef]
8. Wang, Q.; Yue, Y.; Yao, J.; Anderson, J. Adaptive Compensation of Bandwidth Narrowing Effect for Coherent In-Phase Quadrature Transponder through Finite Impulse Response Filter. *Appl. Sci.* **2019**, *9*, 1950. [CrossRef]
9. Zuo, Y.; Zhang, J. A Novel Coding Based Dimming Scheme with Constant Transmission Efficiency in VLC Systems. *Appl. Sci.* **2019**, *9*, 803. [CrossRef]
10. Feng, S.; Feng, H.; Zhou, Y.; Li, B. Low-Complexity Hybrid Optical OFDM with High Spectrum Efficiency for Dimming Compatible VLC System. *Appl. Sci.* **2019**, *9*, 3666. [CrossRef]
11. He, D.; Wang, Q.; Liu, X.; Song, Z.; Zhou, J.; Wang, Z.; Gao, C.; Zhang, T.; Qi, X.; Tan, Y.; et al. Shipborne Acquisition, Tracking, and Pointing Experimental Verifications towards Satellite-to-Sea Laser Communication. *Appl. Sci.* **2019**, *9*, 3940. [CrossRef]

12. Liu, X.; Deng, N. Emerging optical communication technologies for 5G. In *Optical Fiber Telecommunications VII*; Academic Press: Cambridge, MA, USA, 2019; Chapter 17.
13. Katumba, A.; Yin, X.; Dambre, J.; Bienstman, P. Silicon Photonics Neuromorphic Computing and its Application to Telecommunications. In Proceedings of the European Conference on Optical Communication (ECOC), Rome, Italy, 23–27 September 2018.

© 2020 by the authors. Licensee MDPI, Basel, Switzerland. This article is an open access article distributed under the terms and conditions of the Creative Commons Attribution (CC BY) license (http://creativecommons.org/licenses/by/4.0/).

Article

Phase Noise Cancellation in Coherent Communication Systems Using a Radio Frequency Pilot Tone

Tianhua Xu [1,2,3,*], Cenqin Jin [2], Shuqing Zhang [4,*], Gunnar Jacobsen [5,6], Sergei Popov [6], Mark Leeson [2] and Tiegen Liu [1]

1. Key Laboratory of Opto-Electronic Information Technology (MoE), School of Precision Instruments and Opto-Electronics Engineering, Tianjin University, Tianjin 300072, China; tgliu@tju.edu.cn
2. School of Engineering, University of Warwick, Coventry CV4 7AL, UK; Cenqin.Jin@warwick.ac.uk (C.J.); Mark.Leeson@warwick.ac.uk (M.L.)
3. Optical Networks Group, University College London, London WC1E 7JE, UK
4. Department of Optical Engineering, Harbin Institute of Technology, Harbin 150001, China
5. NETLAB, RISE Research Institutes of Sweden, SE-16440 Stockholm, Sweden; gunnar.jacobsen@ri.se
6. Optics and Photonics Group, KTH Royal Institute of Technology, SE-16440 Stockholm, Sweden; sergeip@kth.se
* Correspondence: tianhua.xu@ieee.org (T.X.); sq.zhang@hit.edu.cn (S.Z.)

Received: 26 September 2019; Accepted: 3 November 2019; Published: 5 November 2019

Featured Application: This work is performed for the compensation of the laser phase noise (LPN) and the equalization enhanced phase noise (EEPN) in high-capacity, long-haul, coherent optical fiber networks.

Abstract: Long-haul optical fiber communication employing digital signal processing (DSP)-based dispersion compensation can be distorted by the phenomenon of equalization-enhanced phase noise (EEPN), due to the reciprocities between the dispersion compensation unit and the local oscillator (LO) laser phase noise (LPN). The impact of EEPN scales increases with the increase of the fiber dispersion, laser linewidths, symbol rates, signal bandwidths, and the order of modulation formats. In this work, the phase noise cancellation (PNC) employing a radio frequency (RF) pilot tone in coherent optical transmission systems has been investigated. A 28-Gsym/s QPSK optical transmission system with a significant EEPN has been implemented, where the carrier phase recovery (CPR) was realized using the one-tap normalized least-mean-square (NLMS) estimation and the differential phase detection (DPD), respectively. It is shown that the RF pilot tone can entirely eliminate the LPN and efficiently suppress the EEPN when it is applied prior to the CPR.

Keywords: coherent optical fiber communication; laser phase noise (LPN); carrier phase recovery (CPR); phase noise cancellation (PNC); equalization enhanced phase noise (EEPN); radio frequency (RF) pilot tone

1. Introduction

Long-haul optical communication is seriously deteriorated by transmission impairments, e.g., chromatic dispersion (CD), polarization mode dispersion (PMD), laser phase noise (LPN), and Kerr fiber nonlinearities [1,2]. The combination of coherent detection, digital signal processing (DSP), and advanced modulation formats offers a very promising solution for long-haul, high-capacity optical transmission, to offer great capabilities and flexibilities in the design, deployment, and operation of core telecommunication networks [3–5]. In the reported phase noise cancellation (PNC) methods, the radio frequency (RF) pilot tone scheme was verified to be an efficient approach to remove laser phase fluctuations [6–9]. However, these works only studied the behaviors of RF pilot tones in short-reach systems, where the enhancement effect of fiber dispersion on the LPN was neglected [10,11]. Actually, due to the interplay between the electronic dispersion compensation (EDC) module and the LPN from the local oscillator (LO), a phenomenon of equalization enhanced phase noise (EEPN) is induced and plays a significant role in the carrier phase recovery (CPR) in high-capacity optical communication systems [10–16]. However, so far, no DSP-based CPR has been developed to effectively compensate for EEPN [16–18]. Therefore, it will be of great significance to study the suppression of LPN and EEPN using an RF pilot tone scheme.

In this work, the performance of an orthogonally polarized RF pilot tone scheme is investigated for eliminating both the LPN and the EEPN in long-haul optical transmission systems. A 28-Gsym/s quadrature phase shift keying (QPSK) system is numerically implemented, where the CPR is realized using a one-tap normalized least mean square (NLMS) estimation and a differential phase detection (DPD) scheme, respectively [19–21]. The results show that the LPN can be fully removed and the EEPN can also be effectively suppressed, when the RF pilot tone scheme is applied prior to the CPR.

2. EEPN in Optical Communication Systems

Figure 1 describes the block diagram of a long-haul, coherent optical communication system with EDC and CPR. The LPN from the transmitter (Tx) laser passes through the optical fiber and the dispersion compensation unit, and thus the net CD experienced by the Tx LPN approaches zero. By contrast, the LO LPN goes only through the dispersion compensation unit. Consequently, the LO LPN interplays with the dispersion compensation component in EDC and the interactions will degrade the performance of the optical communication system. This effect is called EEPN [10–13].

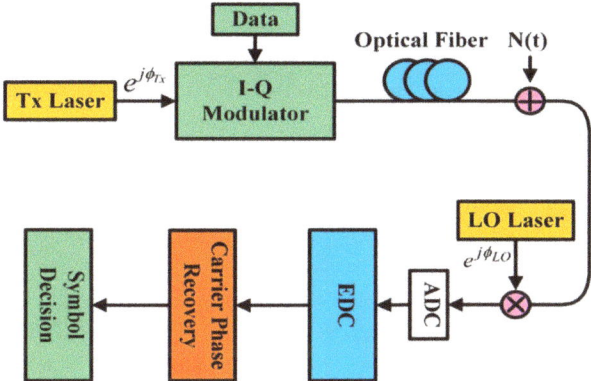

Figure 1. Block diagram of coherent transmission system and equalization-enhanced phase noise (EEPN). ϕ_{Tx}: Tx laser phase noise (LPN), ϕ_{LO}: LO LPN, ADCs: analogue-to-digital converters.

The variance of the EEPN distortion will increase with fiber dispersion, local oscillator laser linewidth, and signal symbol rate. The noise variance of EEPN can be written as follows [10,19]:

$$\sigma_{EEPN}^2 = \frac{\pi \lambda^2}{2c} \cdot \frac{D \cdot L \cdot \Delta f_{LO}}{T_S}, \tag{1}$$

where λ is the central wavelength of the optical carrier, c is the speed of the light in vacuum, L is the length, D is the CD coefficient of the fiber, T_S is the symbol period of the signal, and Δf_{LO} is the 3-dB linewidth of the LO laser.

It is noted that the EEPN evaluation in Equation (1) only works for the static time-domain and frequency-domain EDCs, which involve no phase noise compensation functions [22,23].

3. CPR Using One-Tap NLMS

A one-tap NLMS filter could be effectively applied in the CPR [20], and its tap weight $w(k)$ is expressed using the following equations:

$$w(k+1) = w(k) + \frac{\mu}{|x(k)|^2} x^*(k)e(k), \quad (2)$$

$$e(k) = d(k) - w(k) \cdot x(k), \quad (3)$$

where $x(k)$ is the input signal, k is the symbol index, $d(k)$ is the desired symbol, $e(k)$ is the error between the output and the desired symbols, and μ represents a parameter of the step size.

The schematic of the NLMS CPR is illustrated in Figure 2, which can actually be implemented in a feed-forward structure and can be realized in parallel in the field programmable gate array (FPGA) circuits for a real-time operation [4].

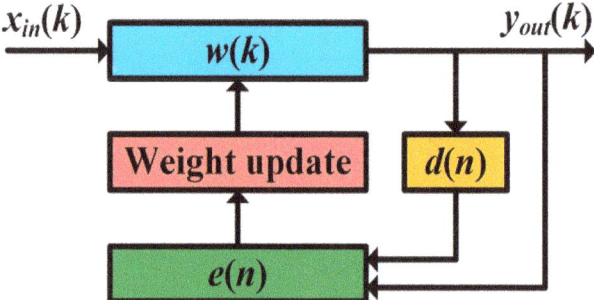

Figure 2. Block diagram of the one-tap normalized least-mean-square (NLMS) carrier phase recovery (CPR).

Similar to the definition of the Tx and the LO, a concept of effective linewidth Δf_{Eff} is employed here to describe the total phase noise in coherent transmission systems considering EEPN [17,19]:

$$\Delta f_{Eff} \approx \frac{\sigma_{Tx}^2 + \sigma_{LO}^2 + \sigma_{EEPN}^2}{2\pi T_S} \quad (4)$$

$$\sigma_{Tx}^2 = 2\pi \Delta f_{Tx} \cdot T_S \quad (5)$$

$$\sigma_{LO}^2 = 2\pi \Delta f_{LO} \cdot T_S \quad (6)$$

where σ_{Tx}^2 and σ_{LO}^2 are the LPN variance of the Tx and LO lasers, respectively, and Δf_{Tx} is the 3-dB linewidth of the Tx laser.

It has been established that the step size μ can be optimized to enhance the performance of NLMS CPR [19]. Therefore, it is significant to find the optimal step size for NLMS CPR, when the EEPN is considered in coherent optical communication systems. Figure 3 shows the optimal step size parameter for various effective linewidths in the 28-Gsym/s QPSK system, applicable to the time-domain and the frequency-domain static dispersion equalizations [22,23]. In all numerical simulations in this paper, the NLMS algorithm is operated with its corresponding optimal value of the step size.

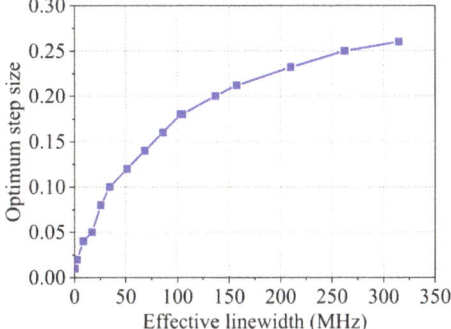

Figure 3. The optimal step size parameter in the NLMS CPR for various effective linewidths.

4. Differential Carrier Phase Recovery

Differential phase detection (DPD) can also be employed for recovering the carrier phase in coherent communication systems. In the DPD scheme, signal data are encoded in and extracted from the phase differences between consecutive transmitted signal symbols [21]. For instance, the phase information of the k-th symbol can be extracted according to the phase difference between the k-th and the (k + 1)-th symbols. Therefore, the CPR error is determined by the phase fluctuation between the k-th and the (k + 1)-th symbols within a symbol period [19]. It is noted that DPD does not require additional computations such as the n-power, the averaging, and the phase unwrapping, compared to other CPR methods [20,24–26]. Therefore, it can be easily implemented in DSP hardware for a real-time operation. The block diagram of DPD CPR is provided in Figure 4.

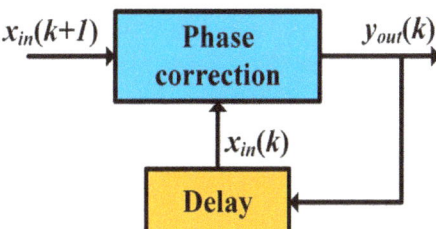

Figure 4. Schematic of the differential phase detection.

5. Transmission Setup with RF Pilot Tone Scheme

Figure 5 depicts a 28-Gsym/s QPSK coherent optical fiber transmission system using an RF pilot tone orthogonally polarized against the transmitted signals. The electrical data were converted into 28-Gsym/s QPSK signals using an in-phase and quadrature (I-Q) optical modulator. The modulated signals occupied one polarization state, and the RF pilot tone was transmitted in the orthogonal polarization state. The signals and the RF pilot tone are integrated into the fiber via a polarization beam combiner (PBC). At the receiver side, both the received signals and the RF pilot tone are mixed with the LO laser, respectively, and are then detected by balanced photodiodes after the 90° hybrid. After that, both the received signals and the RF pilot tone are digitized using 8-bit analogue-to-digital converters (ADCs) at 56 Gsym/s and are then equalized using the EDC module. The signals are further multiplied with a conjugate of the processed RF pilot tone to suppress the impact of both LPN and EEPN. The central wavelengths of both the transmitter and the local oscillator lasers are 1553.6 nm, and the fiber dispersion coefficient is 16 ps·nm^{-1} km^{-1}. Here, we have neglected fiber attenuation,

PMD, and Kerr nonlinearities [27,28]. The EDC is implemented in the frequency domain [23]. The CPR is performed using the one-tap NLMS and DPD methods, respectively.

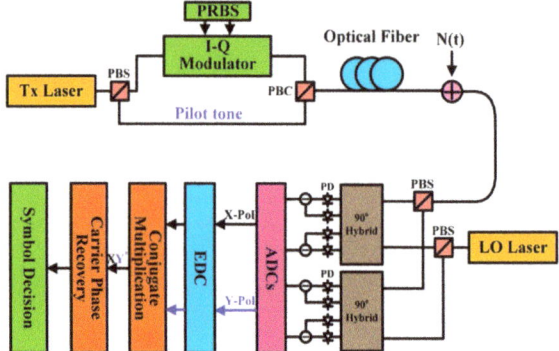

Figure 5. High-speed coherent communication system with an orthogonally-polarized RF pilot tone.

6. Results and Analyses

Figure 6 shows the results of PNC using the RF pilot tone in a 2000 km coherent communication system, where the CPR is realized using the NLMS algorithm. In Figure 6a, the effective linewidth is 170 MHz and there is no EEPN, since the linewidth of the transmitter laser is 170 MHz and the linewidth of the LO laser is 0 Hz. It is found that both the Tx and the LO LPN can be entirely suppressed. In Figure 6b, the linewidths of both the Tx and the LO lasers are 5 MHz. According to Equations (1) and (4), the EEPN is quite significant in such a case, and the effective linewidth is the same (170 MHz) as that in Figure 6a. It is found in Figure 6b that the BER performance improves considerably (around half an order of magnitude in the BER floor, from 8×10^{-4} to 3×10^{-4}) using the RF pilot tone, compared to the case of NLMS CPR only. The results in Figure 6a,b demonstrate the efficiency of the RF pilot tone in compensating for both LPN and EEPN. It can be found that the use of RF pilot tone can entirely mitigate the LPN from both the Tx and the LO lasers. However, it cannot fully compensate for the EEPN in optical fiber transmission systems since the EEPN actually represents a complicated integration of the phase, the amplitude, and the time jitter noise [10,12,16,29].

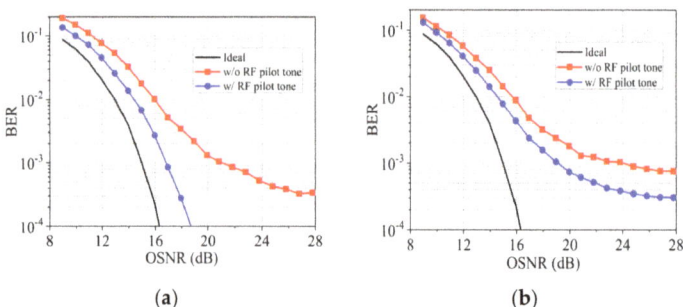

Figure 6. Phase noise cancellation (PNC) results of the 2000 km coherent transmission system (the one-tap NLMS CPR). Ideal: linewidth of both Tx and LO lasers is 0 Hz. w/o: without, w/: with. (**a**) Transmitter: 170 MHz, local oscillator: 0 Hz; (**b**) Transmitter = local oscillator: 5 MHz.

When the CPR is implemented using DPD, the results of the PNC using the RF pilot tone in the same 2000 km coherent system are illustrated in Figure 7. In Figure 7a, the linewidth of the transmitter laser is 170 MHz and the linewidth of the LO laser is 0 Hz, representing an absence of EEPN. It is found that the full suppression of LPN can be performed with the use of the RF pilot tone, compared to the scheme of DPD only. In Figure 7b, the linewidths of both the transmitter and local oscillator lasers are 5 MHz, so that the EEPN is significant again (with an effective linewidth of 170 MHz). Similar to Figure 6b, a considerable improvement of BER performance is also achieved with half an order of magnitude in the BER floor (from 6.5×10^{-4} to 1.5×10^{-4}), when the RF pilot tone scheme is applied prior to the DPD.

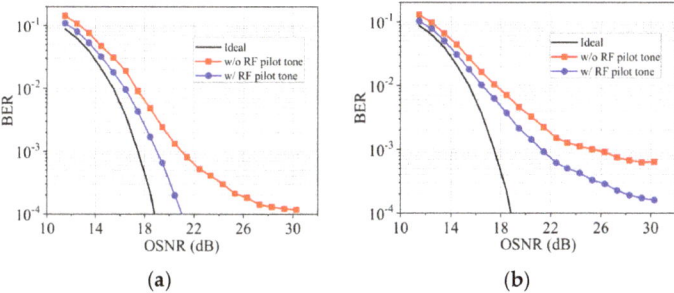

Figure 7. PNC results of the 2000 km coherent transmission system (DPD CPR). Ideal: linewidth of both Tx and LO lasers are 0 Hz. (**a**) Transmitter: 170 MHz, local oscillator: 0 Hz; (**b**) Transmitter = local oscillator: 5 MHz.

7. Discussions

It has to be clarified that the use of the RF pilot tone for the PNC in this work has occupied one polarization state in the fiber, since here we aim to investigate and discuss the efficiency of the use of an RF pilot tone for PNC in a simplified transceiver structure. In fact, the RF pilot tone can also be transmitted using the same polarization state as the optical data signals, which requires a more complicated implementation to perform the generation and recovery of the RF pilot tone [7,30,31]. In such systems, a pilot-carrier vector modulation (PCVM) scheme has been applied, where the transverse magnetic (TM) component is loaded with the signal data in X-polarization and the transverse electric (TE) component is employed to carry both the signal data in Y-polarization and the RF pilot tone [7]. It was reported that the PNC in such PCVM transmission schemes can also provide a good performance for an effective linewidth of up to 30 MHz [7,31]. The performance comparison between such PCVM transmission scheme and our proposed RF pilot tone-PNC scheme will be investigated in future work. In addition, for orthogonal frequency division multiplexing (OFDM) optical transmission systems it will be straightforward to employ one subcarrier as the RF pilot tone within the OFDM spectrum to remove the influence of laser phase noise and EEPN [32].

To study the impact of PMD on the proposed RF pilot tone PNC scheme, numerical simulations using the same 28-Gsym/s QPSK coherent transmission setup have been implemented with a PMD coefficient of $0.1\ ps/\sqrt{km}$. The differential group delay (DGD) and the random rotations between the transmitted data and the RF pilot tone (due to effect of PMD) are mitigated using a constant modulus algorithm CMA equalizer [5,33,34]. The CPR is performed using the one-tap NLMS approach. Figure 8 shows the PNC results of the same 2000 km coherent system, with and without including the impact of PMD. Similarly, the linewidth of the Tx laser is 170 MHz and the linewidth of the LO laser is 0 Hz in Figure 8a, and the linewidths of both the Tx and the LO lasers are 5 MHz in Figure 8b. Both scenarios indicate an effective linewidth of 170 MHz, and there is no EEPN in Figure 8a. It can be clearly found that, for both transmission scenarios, the impact of PMD can be fully suppressed using the CMA equalizer and the performance of the RF pilot tone-based PNC (both with and without EEPN) will not be affected due to the introduction of PMD.

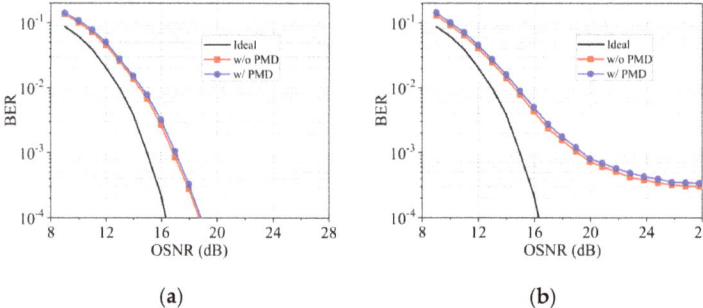

Figure 8. PNC performance of the 2000 km coherent transmission systems with and without PMD, when the NLMS-CPR is applied. Ideal: linewidth of both Tx and LO lasers are 0 Hz. (**a**) Transmitter: 170 MHz, local oscillator: 0 Hz; (**b**) Transmitter = local oscillator: 5 MHz.

8. Conclusions

In this work, an orthogonally polarized RF pilot tone is investigated to suppress the LPN and EEPN in long-haul coherent optical fiber communication systems. A 28-Gsym/s QPSK optical transmission system is numerically implemented, where the CPR is performed using the one-tap NLMS and DPD methods, respectively. Our results demonstrate that the LPN in the communication system can be completely eliminated and the EEPN can also be effectively suppressed, when the RF pilot tone scheme is applied prior to the CPR.

Author Contributions: Theoretical analyses and numerical simulations were implemented by T.X. The paper was mainly written by T.X., C.J. wrote some parts and edited the manuscript. S.Z. and M.L. made modifications. All authors reviewed and discussed the paper.

Funding: This work is in parts supported by the EU H2020 RISE Grant DAWN4IoE (778305) and the UK EPSRC Program Grant TRANSNET (EP/R035342/1).

Conflicts of Interest: The authors declare no conflict of interest.

References

1. Kaminow, I.; Li, T.; Willner, A.E. *Optical Fiber Telecommunications VIA: Components and Subsystems*, 6th ed.; Academic Press Publisher: Oxford, UK, 2013.
2. Ip, E.; Lau, A.P.T.; Barros, D.J.F.; Kahn, J.M. Coherent detection in optical fiber systems. *Opt. Express* **2008**, *16*, 753–791. [CrossRef] [PubMed]
3. Winzer, P.J.; Essiambre, R.J. Advanced modulation formats for high-capacity optical transport networks. *J. Lightwave Technol.* **2006**, *24*, 4711–4728. [CrossRef]
4. Savory, S.J. Digital coherent optical receivers: Algorithms and subsystems. *IEEE J. Sel. Top. Quantum Electron.* **2010**, *16*, 1164–1179. [CrossRef]
5. Zhao, J.; Liu, Y.; Xu, T. Advanced DSP for coherent optical fiber communication. *App. Sci.* **2019**, *9*, 4192. [CrossRef]
6. Mori, Y.; Zhang, C.; Igarashi, K.; Katoh, K.; Kikuchi, K. Unrepeated 200 km transmission of 40-Gbit/s 16-QAM signals using digital coherent receiver. *Opt. Express* **2009**, *17*, 1435–1441. [CrossRef]
7. Nakamura, M.; Kamio, Y.; Miyazaki, T. Pilot-carrier based linewidth-tolerant 8PSK self-homodyne using only one modulator. In Proceedings of the 33rd European Conference and Exhibition of Optical Communication, Berlin, Germany, 16–20 September 2007.
8. Cai, Y.; Gao, X.; Ling, Y.; Xu, B.; Qiu, K. RF pilot tone phase noise cancellation based on DD-MZM SSB modulation for optical heterodyne RoF link. *Opt. Commun.* **2020**, *454*, 124502. [CrossRef]
9. Jacobsen, G.; Xu, T.; Popov, S.; Li, J.; Friberg, A.T.; Zhang, Y. Receiver implemented RF pilot tone phase noise mitigation in coherent optical nPSK and nQAM systems. *Opt. Express* **2011**, *19*, 14487–14494. [CrossRef]

10. Shieh, W.; Ho, K.P. Equalization-enhanced phase noise for coherent detection systems using electronic digital signal processing. *Opt. Express* **2008**, *16*, 15718–15727. [CrossRef]
11. Xie, C. Local oscillator phase noise induced penalties in optical coherent detection systems using electronic chromatic dispersion compensation. In Proceedings of the 2009 Conference on Optical Fiber Communication-Incudes post deadline papers, San Diego, CA, USA, 29 May 2009.
12. Lau, A.P.T.; Shen, T.S.R.; Shieh, W.; Ho, K.P. Equalization-enhanced phase noise for 100 Gb/s transmission and beyond with coherent detection. *Opt. Express* **2010**, *18*, 17239–17251. [CrossRef]
13. Xu, T.; Liga, G.; Lavery, D.; Thomsen, B.C.; Savory, S.J.; Killey, R.I.; Bayvel, P. Equalization enhanced phase noise in Nyquist-spaced superchannel transmission systems using multi-channel digital back-propagation. *Sci. Rep.* **2015**, *5*. [CrossRef]
14. Zhuge, Q.; Xu, X.; El-Sahn, Z.A.; Mousa-Pasandi, M.E.; Morsy-Osman, M.; Chagnon, M.; Qiu, M.; Plant, D.V. Experimental investigation of the equalization-enhanced phase noise in long haul 56 Gbaud DP-QPSK systems. *Opt. Express* **2012**, *20*, 13841–13846. [CrossRef] [PubMed]
15. Fatadin, I.; Savory, S.J. Impact of phase to amplitude noise conversion in coherent optical systems with digital dispersion compensation. *Opt. Express* **2010**, *18*, 16273–16278. [CrossRef] [PubMed]
16. Kakkar, A.; Navarro, J.R.; Schatz, R.; Louchet, H.; Pang, X.; Ozolins, O.; Jacobsen, G.; Popov, S. Comprehensive study of equalization-enhanced phase noise in coherent optical systems. *J. Lightwave Technol.* **2015**, *33*, 4834–4841. [CrossRef]
17. Xu, T.; Jacobsen, G.; Popov, S.; Li, J.; Friberg, A.T.; Zhang, Y. Carrier phase estimation methods in coherent optical transmission systems influenced by equalization enhanced phase noise. *Opt. Commun.* **2013**, *293*, 54–60. [CrossRef]
18. Colavolpe, G.; Foggi, T.; Forestieri, E.; Secondini, M. Impact of phase noise and compensation techniques in coherent optical systems. *J. Lightwave Technol.* **2011**, *29*, 2790–2800. [CrossRef]
19. Xu, T.; Jacobsen, G.; Popov, S.; Li, J.; Friberg, A.T.; Zhang, Y. Analytical estimation of phase noise influence in coherent transmission system with digital dispersion equalization. *Opt. Express* **2011**, *19*, 7756–7768. [CrossRef]
20. Xu, T.; Jacobsen, G.; Popov, S.; Li, J.; Wang, K.; Friberg, A.T. Normalized LMS digital filter for chromatic dispersion equalization in 112-Gbit/s PDM-QPSK coherent optical transmission system. *Opt. Commun.* **2010**, *283*, 963–967. [CrossRef]
21. Fatadin, I.; Ives, D.; Savory, S.J. Differential carrier phase recovery for QPSK optical coherent systems with integrated tunable lasers. *Opt. Express* **2013**, *21*, 10166–10171. [CrossRef]
22. Savory, S.J. Digital filters for coherent optical receivers. *Opt. Express* **2008**, *16*, 804–817. [CrossRef]
23. Kudo, R.; Kobayashi, T.; Ishihara, K.; Takatori, Y.; Sano, A.; Miyamoto, Y. Coherent optical single carrier transmission using overlap frequency domain equalization for long-haul optical systems. *J. Lightwave Technol.* **2009**, *27*, 3721–3728. [CrossRef]
24. Jacobsen, G.; Xu, T.; Popov, S.; Sergeyev, S. Study of EEPN mitigation using modified RF pilot and Viterbi-Viterbi based phase noise compensation. *Opt. Express* **2013**, *21*, 12351–12362. [CrossRef] [PubMed]
25. Ly-Gagnon, D.S.; Tsukamoto, S.; Katoh, K.; Kikuchi, K. Coherent detection of optical quadrature phase-shift keying signals with carrier phase estimation. *J. Lightwave Technol.* **2006**, *24*, 12–21. [CrossRef]
26. Viterbi, A.J.; Viterbi, A.M. Nonlinear estimation of PSK-modulated carrier phase with application to burst digital transmission. *IEEE Trans. Inf. Theory* **1983**, *29*, 543–551. [CrossRef]
27. Liga, G.; Xu, T.; Alvarado, A.; Killey, R.I.; Bayvel, P. On the performance of multichannel digital backpropagation in high-capacity long-haul optical transmission. *Opt. Express* **2014**, *22*, 30053–30062. [CrossRef] [PubMed]
28. Maher, R.; Xu, T.; Galdion, L.; Sato, M. Spectrally shaped DP-16QAM super-channel transmission with multi-channel digital back propagation. *Sci. Rep.* **2015**, *5*, 08214. [CrossRef] [PubMed]
29. Ho, K.P.; Lau, A.P.T.; Shieh, W. Equalization-enhanced phase noise induced timing jitter. *Opt. Lett.* **2011**, *36*, 585–587. [CrossRef]
30. Nakamura, M.; Kamio, Y.; Miyazaki, T. Linewidth-tolerant real-time 40-Gbit/s 16-QAM self-homodyne detection using a pilot carrier and ISI suppression based on electronic digital processing. *Opt. Lett.* **2010**, *35*, 13–15. [CrossRef]
31. Nakamura, M.; Kamio, Y.; Miyazaki, T. Linewidth-tolerant real-time 10-Gbit/s 16-QAM transmission using a pilot-carrier based phase-noise cancelling technique. *Opt. Express* **2008**, *16*, 10611–10616. [CrossRef]

32. Randel, S.; Adhikari, S.; Jansen, S.L. Analysis of RF-pilot-based phase noise compensation for coherent optical OFDM systems. *IEEE Photon. Technol. Lett.* **2010**, *22*, 1288–1290. [CrossRef]
33. Ip, E.; Kahn, J.M. Digital equalization of chromatic dispersion and polarization mode dispersion. *J. Lightwave Technol.* **2007**, *25*, 2033–2043. [CrossRef]
34. Liga, G.; Czegledi, C.B.; Xu, T.; Agrell, E.; Killey, R.I.; Bayvel, P. Ultra-wideband nonlinearity compensation performance in the presence of PMD. In Proceedings of the 42nd European Conference on Optical Communication, Dusseldorf, Germany, 5 December 2016.

 © 2019 by the authors. Licensee MDPI, Basel, Switzerland. This article is an open access article distributed under the terms and conditions of the Creative Commons Attribution (CC BY) license (http://creativecommons.org/licenses/by/4.0/).

Article

400GbE Technology Demonstration Using CFP8 Pluggable Modules

Yang Yue *, Qiang Wang, Jian Yao, Jason O'Neil, Daniel Pudvay and Jon Anderson

Juniper Networks, 1133 Innovation Way, Sunnyvale, CA 94089, USA; qiwang.thresh@gmail.com (Q.W.); jianyao@juniper.net (J.Y.); joneil@juniper.net (J.O.); dpudvay@juniper.net (D.P.); joanderson@juniper.net (J.A.)
* Correspondence: yyue@juniper.net; Tel.: +1-408-745-2000

Received: 3 October 2018; Accepted: 23 October 2018; Published: 25 October 2018

Abstract: In this article, we first review the current status of 400GBASE client-side optics standards and multi-source agreements (MSAs). We then compare different form factors for 400GE modules, including CFP8, OSFP and QSFP-DD. The essential techniques to implement 400GE, such as pulse amplitude modulation (PAM4), forward error correction (FEC) and a continuous time-domain linear equalizer (CTLE), are discussed. A 400GE physical interface card (PIC) in Juniper's PTX5000 platform has been developed, conforming to the latest IEEE802.3bs standard. To validate the PIC's performance, a commercial optical network tester (ONT) and the PIC are optically interconnected through two CFP8-LR8 modules. The CFP8-LR8 module utilizes eight optical wavelengths through coarse wavelength division multiplexing (CWDM). Each wavelength carries 50 Gb/s PAM4 signal. The signal transmits through 10 km of single mode fiber (SMF). The ONT generates framed 400GE signal and sends it to the PIC through the first CFP8 module. The PIC recovers the signal, performs an internal loopback, and sends 400GE signal back to the ONT through the second CFP8 module. The optical spectrum, eye diagram, receiver sensitivity, long time soaking results, and internal digital diagnosis monitoring (DDM) result are fully characterized. The pre-FEC bit error rate (BER) is well below the KP4 FEC threshold of 2.2×10^{-4}. After KP4 FEC, error-free performance over 30 km of SMF is achieved. In this way, we demonstrate both the interoperation between the PIC and the ONT, as well as the interoperation between the two CFP8 modules. This demonstration represents the successful implementation of the 400GE interface in the core IP/MPLS router.

Keywords: optical communication; fiber optics; client-side optics; 400G Ethernet; CFP8-LR8 transceiver

1. Introduction

Processing information and transmitting information are two fundamental functions in communication networks [1]. From a quantum physics perspective, particles can be classified as fermions or bosons. Electrons follow Fermi-Dirac statistics, while Bose-Einstein statistics apply to photons. Due to strong interaction, electrons are ideal for processing information. As there is minimal interaction between bosons, photons are ideal for transmitting information in different degrees of freedom including wavelength, time, amplitude, phase, polarization, mode, and space [2,3].

Figure 1 illustrates the architecture of a core router, which is widely used in today's communication networks. A typical router includes a routing engine (RE), routing control board (RCB), physical interface card (PIC), flexible PIC concentrator (FPC), and switch interface board (SIB). Information processing in the communication network is mainly performed by these functional units, which is in the electron domain. Moreover, some of Juniper's packet optics products on the PTX platform are also shown in Figure 1. Information transmission is mainly performed by these functional units, which is in the photon domain. The integrated photonics line card (IPLC) tightly couples a 1 × 2 wavelength selective switching (WSS), embedded bi-directional switch gain amplifiers, optical multiplexer/de-multiplexers, and optical supervisory channel for optical management into a single package that supports up to 64 ITU-T C-band wavelengths at 50 GHz spacing via an IPLC expansion line card. P3-15-U-QSFP28 and P2-PTX-5-100G-WDM are PTX Series PICs that support 15 QSFP28 and 5 CFP2 modules.

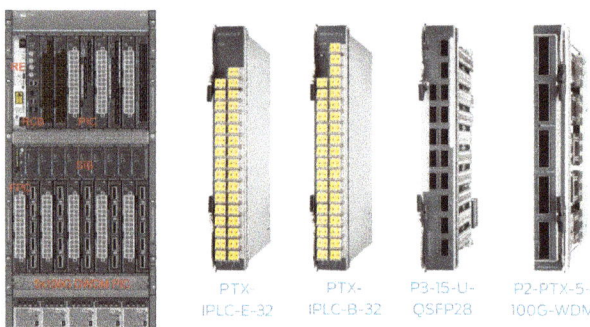

Figure 1. Typical architecture of core router and Juniper packet optical product. RE: routing engine; RCB: routing control board; PIC: physical interface card; FPC: Flexible PIC concentrator; SIB: switch interface board.

In 2017, standardization of the 400 Gigabit Ethernet (400GE) was ratified by the IEEE P802.3bs Task Force [4]. This paves the way for deployment of 400GE in the network. It is expected that the demand for 400GE will grow rapidly over the next couple of years [5]. Recently, multiple industrial line-side and client-side interoperability trials have been successfully demonstrated using 400GE CFP8 pluggable optical modules [6–8]. These trials demonstrate that the ecosystem for the 400GE application is mature. In this paper, we present a successful demonstration of the 400GE physical interface card (PIC) integrated in the core internet protocol/multi-protocol label switching (IP/MPLS) router, with the CFP8-LR8 modules acting as the optical front end.

IEEE defines the optical and electrical parameters for the physical media dependent (PMD) to guarantee interoperability between PMDs. The physical implementations of PMDs are defined by multi-source agreement (MSA) instead. Currently, there are still multiple MSAs defining pluggable optical modules. Figure 2 below shows the comparison between different form factors defined by multiple MSAs [9–16]. The most notable form factors for 400GE defined by MSAs are QSFP-DD, OSFP and CFP8. Currently the industry is converging behind QSFP-DD due to the high port density on the front panel, backward compatibility with QSFP28/QSFP+, and large ecosystem.

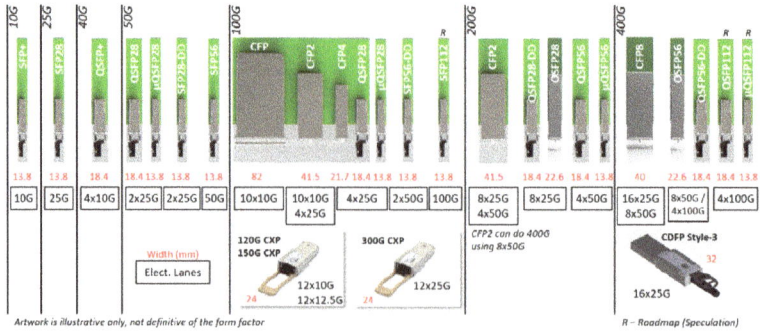

Figure 2. Form factors and electrical lanes for pluggable optical modules.

On-off-keying (OOK) has been the modulation format for 40 Gigabit Ethernet (40GE) and 100 Gigabit Ethernet (100GE). The IEEE 802.3bs has selected 4-level pulse amplitude modulation (PAM4) for 400GE transmission. Compared with OOK, the required baud rate of PAM4 could be reduced by a factor of two when keeping the desired net data rate constant. Thus, the transceiver implements PAM4 with fewer optical lanes/lower bandwidth components. This leads to lower cost, smaller power consumption and a denser footprint.

The following variants are defined in 803.bs: 400GBASE-DR4, 50 gigabit per second (Gb/s) 4 parallel single mode fiber transmission (PSM4) up to 500 m; 400GBASE-FR8, 25 Gb/s 8-λ wavelength division multiplexing (WDM) transmission up to 2 km; and, 400GBASE-LR8, 25 Gb/s 8-λ coarse WDM transmission up to 10 km. All interfaces use a 1300 nm window with minimum chromatic dispersion and a low-cost transceiver. For 8-λ WDM transmission, the assignment of the optical wavelength on the coarse wavelength division multiplexing (CWDM) grid removes the requirement of precise temperature control for the laser diode.

To improve sensitivity and increase the transmission distance, forward error correction (FEC) is widely used in today's client-side optics interface. IEEE has approved two FEC coding schemes, KR4 and KP4. KR4-FEC utilizes Reed-Solomon coding (528, 514). The net coding gain is 5.3 dB and the bit error rate (BER) threshold for the uncorrected code word (UCW) is 2.1×10^{-5}. This is widely used in the non-return-to-zero (NRZ) OOK modulation format. However, with the same symbol rate, the sensitivity penalty of PAM4 over NRZ OOK is $10 \times \log_{10} (1/3)$ = 4.77 dB. In practice, there is further performance degradation due to the system nonlinearity. In order to close the link budget, KP4-FEC defined in IEEE 802.3bs clause 91, is widely used in PAM4 transmission [17]. KP4-FEC utilizes Reed-Solomon coding (544, 514). The net coding gain is 6.4 dB and the BER threshold for the UCW is 2.2×10^{-4}.

Figure 3 illustrates PAM4 eye diagrams without and with pre-compensation. As one can see, especially for the high baud rate PAM4 signal, its eye can easily be closed due to system bandwidth limitation. Consequently, it is critically important to implement pre-compensation techniques to improve the signal quality and open up the PAM4 eye. A continuous time-domain linear equalizer (CTLE) has been widely used to equalize the frequency-dependent loss for the electrical interface between the packet forwarding engine (PFE) and the pluggable optical module. CTLE is essentially a high-pass filter which inverts the frequency response of the trace on the printed circuit board (PCB). The typical response of CTLE is plotted in Figure 4. In Reference [4], the emphasis of the CTLE is defined in 1 dB steps. However, 0.5 dB steps may be required in large-scale deployment to improve granularity.

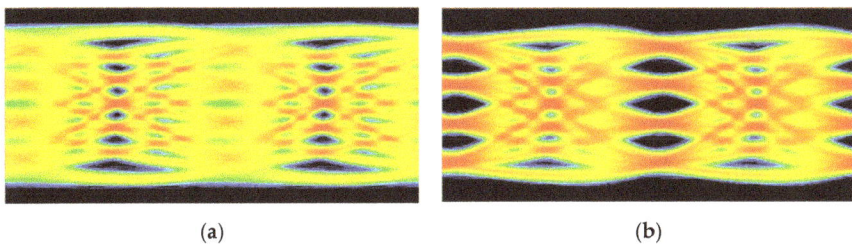

Figure 3. (**a**) Closed 4-level pulse amplitude modulation (PAM4) eye without pre-compensation, (**b**) open PAM4 eye with pre-compensation.

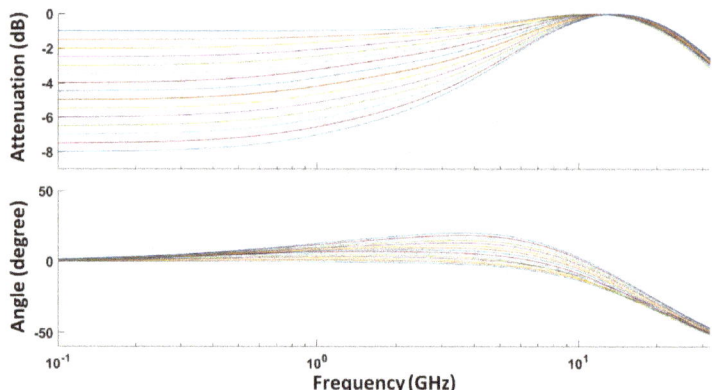

Figure 4. Typical response curve of the continuous time-domain linear equalizer (CTLE) as defined in Reference [4] in 0.5 dB step size.

2. 400GE Pluggable Module and Host Card

Figure 5 below shows the 400GE CFP8-LR8 pluggable optical transceiver and the block diagram of the physical interface card (PIC) hosting the 400GE pluggable module. The total data rate is 425 Gbit/s, due to the overhead of the FEC block and the 64B/66B mapping in the physical coding sublayer (PCS). The data are carried with 8 optical channels modulated using PAM4 signal running at 53.125 Gbit/s (8 × 50G PAM4 WDM optical channels). The electrical interfaces (CDAUI-16) are 16 pairs of differential lanes running at 26.5625 Gbit/s (16 × 25G electrical NRZ signals). The CFP8-LR8 uses the distributed feedback laser (DFB) as the transmitter and the PIN photodiode (p-type, intrinsic, n-type photodiode) as the receiver. The total power consumption is less than 16 W. Within the CFP8-LR8 module, a gearbox integrated circuit (IC) multiplexes two electrical lanes of NRZ signal into one optical lane of pulse amplitude modulation (PAM) signal. Gray mapping is implemented so that there is only a one bit difference between the adjacent levels to minimize the BER.

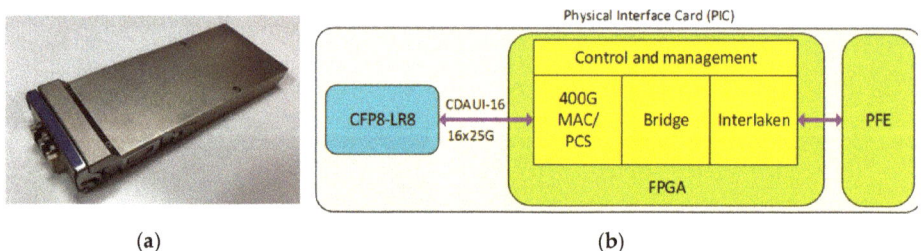

(a) (b)

Figure 5. (a) Picture of 400GE CFP8-LR8 pluggable optical transceiver; (b) configuration of physical interface card (PIC). MAC: media access control, PCS: physical coding sublayer, FPGA: field programmable gate array, PFE: packet forwarding engine, CDAUI-16: 400G attachment unit interface, running at 16 lanes with 26.5625 Gb/s per lane.

Juniper's PIC hosts the PFE which forwards the IP packet based on the routing table generated from the routing engine (RE). The additional interfaces, like the bridge, the media access control (MAC) and the PCS layer are implemented with a field programmable gate array (FPGA). The control and management function is implemented in FPGA as well. One PIC can host three pluggable optical modules, giving a total capacity of the PIC of 1.2 Tbit/s.

Figure 6 shows the experimental setup to demonstrate the interoperation of CFP8-LR8 modules. We use an optical network tester (ONT, Viavi) to generate a CDAUI-16 signal. One CFP8-LR8 module is inserted into the ONT to perform electrical-to-optical conversion. The optical signal goes through a variable optical attenuator (VOA) and is received by a second CFP8-LR8 module, which is hosted by a Juniper's PIC. Within the PIC, the 400GE signal is electrically looped back to the transmitter of the second CFP8-LR8 module. The signal either bypasses or goes through a single mode fiber (SMF) spool. The signal is then received by the CFP8-LR8 from the first module, and the ONT counts the BER.

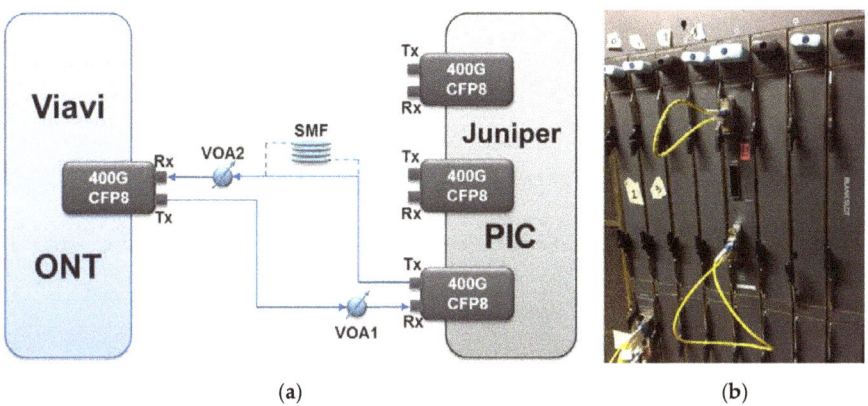

Figure 6. (a) Experimental demonstration of the interoperation of CFP8-LR8 modules; (b) two CFP8-LR8 modules in Juniper's PIC and PTX5000 chassis.

3. Transmitter Performance

Figure 7 below shows the optical eye diagram obtained through Keysight's digital communication analyzer (DCA). The CDAUI-16 electrical signal is generated by a Viavi optical network tester (ONT) and passed through a pluggable connector. A clock signal derived from the ONT drives the DCA to obtain these optical eye diagrams. As seen, a clear 4-level eye diagram with wide eye opening indicates a good signal to noise ratio (SNR). One can also notice that there is a misalignment in the center of the eye, due to the nonlinear response of the modulator.

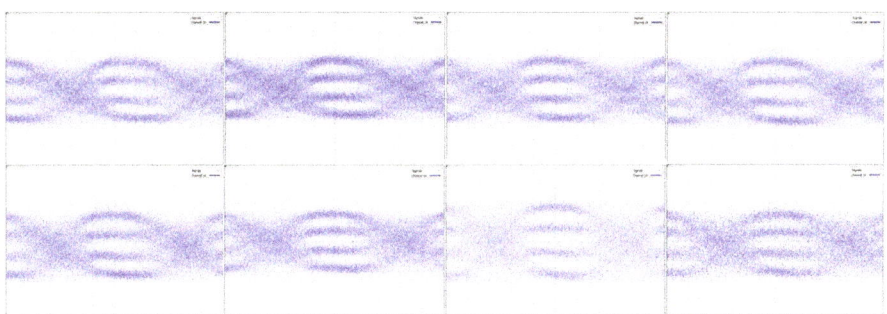

Figure 7. Optical eye diagrams of eight wavelength division multiplexing (WDM) channels using PAM4 modulation format. Top: channels 1 to 4; bottom: channels 5 to 8.

Figure 8 shows the optical spectrum from the CFP8-LR8 modules. The optical wavelengths are around 1310 nm so that the influence of chromatic dispersion is minimized. The wavelength assignment of these eight optical channels follows the CWDM designation. This facilitates the interoperation between the CFP8 modules. Also, the measured side mode suppression ratio (SMSR) is well above 40 dB, indicating the excellent jitter performance for the optical modules.

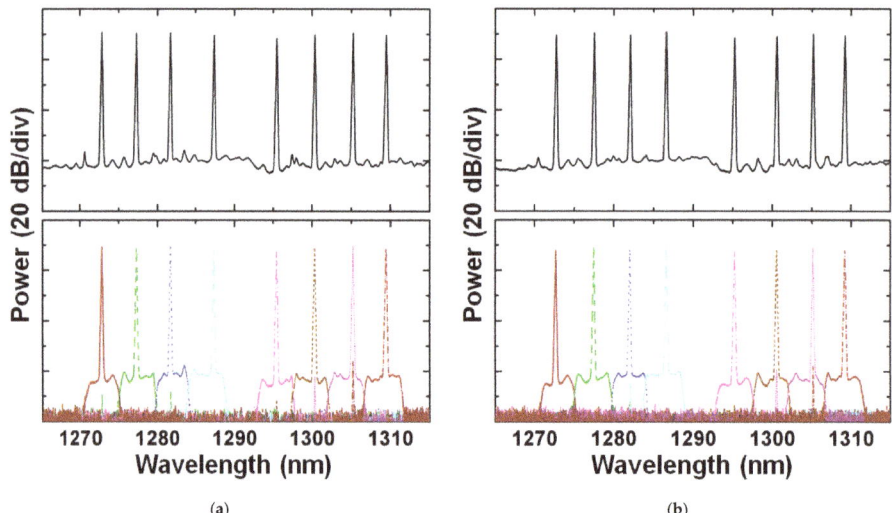

Figure 8. Optical spectra of two CFP8-LR8 modules: (**a**) first sample module; (**b**) second sample module. Top: spectrum at the output of the transmitter; Bottom: spectra of the individual lanes after going through a de-multiplexer.

4. Link Performance

In this section, we first study the system performance in a back-to-back scenario, and then extend our experiment to tens of kilometers of fiber transmission.

4.1. Back-to-Back Performance

We first measure the performance of the 400GE CFP8-LR8 module in a back-to-back scenario. Here we bypass the SMF spool as shown in Figure 9. We first adjust the VOA1 to test the interoperation between the transmitter (Tx) of the first module to the receiver (Rx) of the second module. Meanwhile, the attenuation of VOA2 is set to zero so that there is minimum error introduced on the backward path from the PIC to the ONT. As seen, even with 14 dB attenuation between the transmitter and the receiver, the BER is still below the FEC's UCW threshold of 2.2×10^{-4}. For the BER performance of 16 ONT electrical lanes, we note that some lanes perform better than others. This is potentially because of the non-ideal combination of the PAM4 signal from two individual NRZ OOK signals.

Next, we adjust the VOA2 to test the interoperation between the Tx of the second module and the Rx of the first module. Meanwhile, the attenuation of VOA1 is set to zero so that there is minimum error introduced on the forward path from the ONT to the PIC. As seen in Figure 10, even with 12 dB attenuation between the transmitter and the receiver, the BER is still below the FEC's UCW threshold of 2.2×10^{-4}.

We also measure the receiver optical power (ROP) at the different values of attenuation. There is a difference of 2.3 dB between the launching powers of eight optical lanes. At 12 dB attenuation, the average ROP is -12.4 dBm.

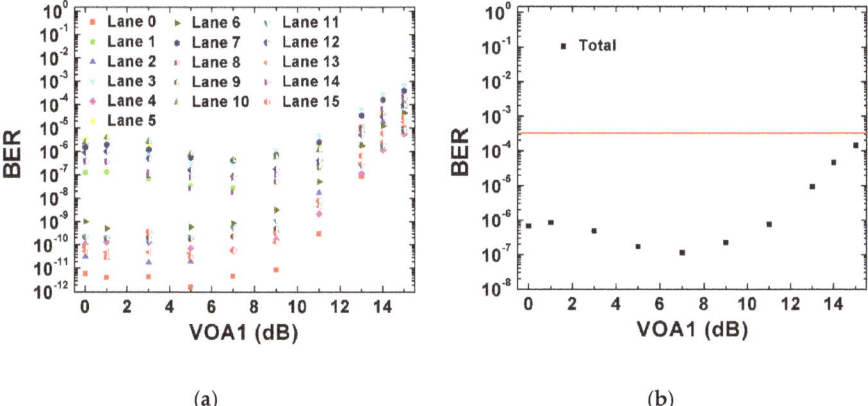

(a) (b)

Figure 9. Back-to-back performance for interoperation of two CFP8-LR8 modules. The signal is transmitted from the transmitter (Tx) of the first module to the receiver (Rx) of the second module: (**a**) Bit error rate (BER) vs. VOA1's attenuation for 16 individual electrical lanes; (**b**) total BER vs. VOA1's attenuation. The threshold for the uncorrected code word (UCW) is shown as the red line.

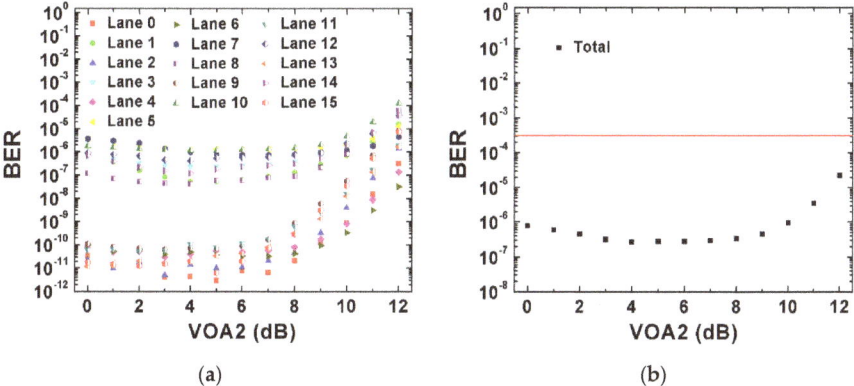

(a) (b)

Figure 10. Back-to-back performance for interoperation of two CFP8-LR8 modules. The signal is transmitted from the Tx of the second module to the Rx of the first module: (**a**) BER versus VOA2's attenuation for 16 individual electrical lanes; (**b**) total BER vs. VOA2's attenuation. The threshold for UCW is shown as the red line.

4.2. Fiber Transmission

Furthermore, we measure the performance of the 400GE CFP8-LR8 module with the single mode fiber. The attenuation of VOA1 is set to zero so that there is minimum error introduced on the forward path from the ONT to the PIC. We adjust the fiber length and measure the BER. The result is shown in Figure 11. As seen, even with 30 km SMF between the transmitter and the receiver, the BER is still below the FEC's UCW threshold of 2.2×10^{-4}. This clearly demonstrates 400G Ethernet interoperation using two CFP8-LR8 modules.

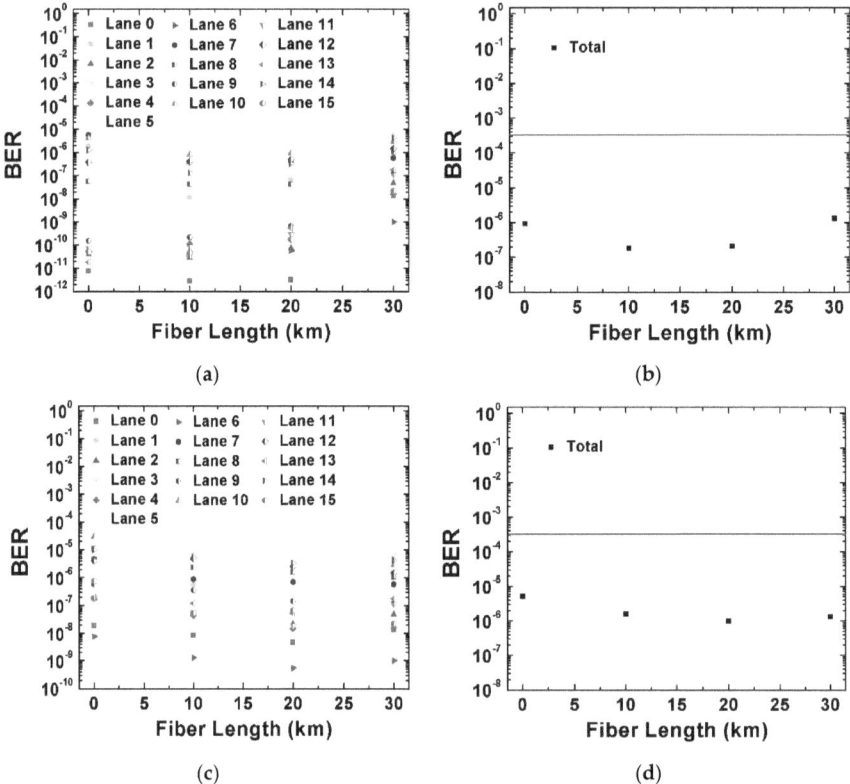

Figure 11. Interoperation of two CFP8-LR8 modules through single mode fiber (SMF) spool. The signal is transmitted from the Tx of the first module to the Rx of the second module: (**a**) BER vs. fiber length for 16 individual electrical lanes; (**b**) total BER vs. fiber length. The signal is transmitted from the Tx of the second module to the Rx of the first module: (**c**) BER versus VOA2's attenuation for 16 individual electrical lanes; (**d**) total BER vs. VOA2's attenuation. The threshold for the UCW is shown as the red line.

5. Conclusions

In this article, we review the current status of the 400G Ethernet standard. The essential techniques to implement 400G client-side optics, like PAM4, FEC and CTLE, are discussed. The PIC in Juniper's PTX5000 platform has been developed with the CFP8-LR8 modules being the optical front-end. The optical signal transmits through 30 km SMF. The pre-FEC BER is well below the threshold for UCW and there is no post-FEC error. This demonstration represents the successful implementation of a 400G Ethernet interface in the core IP/MPLS router using the CFP8-LR8 pluggable modules.

Author Contributions: Conceptualization, Y.Y. and J.A.; methodology, Y.Y.; software, Y.Y. and Q.W.; validation, Y.Y.; formal analysis, Y.Y. and Q.W.; investigation, Y.Y., Q.W. and J.Y.; resources, Y.Y., Q.W., J.Y., J.O. and D.P.; data curation, Y.Y.; writing—original draft preparation, Y.Y. and Q.W.; writing—review and editing, Y.Y., Q.W., J.Y., J.O., D.P. and J.A.; visualization, Y.Y. and Q.W.; supervision, J.A.

Funding: This research received no external funding.

Acknowledgments: The authors gratefully acknowledge the vigorous encouragement and strong support for innovation from Domenico Di Mola at Juniper Networks.

Conflicts of Interest: The authors declare no conflicts of interest.

References

1. Kaminow, I.; Li, T.; Willner, A.E. *Optical Fiber Telecommunications*, 6th ed.; Academic Press: San Diego, CA, USA, 2013; ISBN 9780123969606.
2. Bozinovic, N.; Yue, Y.; Ren, Y.; Tur, M.; Kristensen, P.; Huang, H.; Willner, A.; Ramachandran, S. Terabit-scale orbital angular momentum mode division multiplexing in fibers. *Science* **2013**, *340*, 1545–1548. [CrossRef] [PubMed]
3. Winzer, P.J. Making spatial multiplexing a reality. *Nat. Photonics* **2014**, *8*, 345–348. [CrossRef]
4. IEEE P802.3bs 400 Gb/s Ethernet Task Force. Available online: www.ieee802.org/3/bs/ (accessed on 1 September 2018).
5. Rokkas, T.; Neokosmidis, I.; Tomkos, I. Cost and Power Consumption Comparison of 400 Gbps Intra-Datacenter Transceiver Modules. In Proceedings of the 2018 International Conference on Transparent Optical Networks (ICTON), Bucharest, Romania, 1–5 July 2018.
6. Nelson, L.E.; Zhang, G.; Padi, N.; Skolnick, C.; Benson, K.; Kaylor, T.; Iwamatsu, S.; Inderst, R.; Marques, F.; Fonseca, D.; et al. SDN-Controlled 400GbE end-to-end service using a CFP8 client over a deployed, commercial flexible ROADM system. In Proceedings of the 2017 Optical Fiber Communications Conference and Exposition (OFC), Los Angeles, CA, USA, 19–23 March 2017.
7. Birk, M.; Nelson, L.E.; Zhang, G.; Cole, C.; Yu, C.; Akashi, M.; Hiramoto, K.; Fu, X.; Brooks, P.; Schubert, A.; et al. First 400GBASE-LR8 interoperability using CFP8 modules. In Proceedings of the 2017 Optical Fiber Communications Conference and Exposition (OFC), Los Angeles, CA, USA, 19–23 March 2017.
8. Nelson, L.E. Advances in 400 Gigabit Ethernet Field Trials. In Proceedings of the 2018 Optical Fiber Communications Conference and Exposition (OFC), San Diego, CA, USA, 11–15 March 2018.
9. CFP-MSA. Available online: http://www.cfp-msa.org/ (accessed on 1 September 2018).
10. SFF Committee. Available online: http://www.sffcommittee.com/ie/ (accessed on 1 September 2018).
11. OSFP. Available online: http://osfpmsa.org/index.html (accessed on 1 September 2018).
12. QSFP-DD. Available online: http://www.qsfp-dd.com/ (accessed on 1 September 2018).
13. OIF: Optical Internetworking Forum. Available online: http://www.oiforum.com/ (accessed on 1 September 2018).
14. Cole, C. Beyond 100G client optics. *IEEE Commun. Mag.* **2012**, *50*, s58–s66. [CrossRef]
15. Yue, Y.; Wang, Q.; Maki, J.J.; Zhang, B.; O'Neil, J.; Vovan, A.; Anderson, J. Latest Industry Trend in Pluggable Optics. In Proceedings of the 2017 International Conference on Optical Communications and Networks (ICOCN), Wuzhen, China, 7–10 August 2017.
16. Isono, H. Latest standardization trends for client and networking optical transceivers and its future directions. In Proceedings of the 2018 optoelectronics, photonic materials and devices conference (SPIE OPTO), San Francisco, CA, USA, 27 January–1 February 2018.
17. Chagnon, M.; Lessard, S.; Plant, D.V. 336 Gb/s in Direct Detection below KP4 FEC Threshold for Intra Data Center Applications. *IEEE Photonics Technol. Lett.* **2016**, *28*, 2233–2236. [CrossRef]

© 2018 by the authors. Licensee MDPI, Basel, Switzerland. This article is an open access article distributed under the terms and conditions of the Creative Commons Attribution (CC BY) license (http://creativecommons.org/licenses/by/4.0/).

Article

Power and Signal-to-Noise Ratio Optimization in Mesh-Based Hybrid Optical Network-on-Chip Using Semiconductor Optical Amplifiers

Jun Yeong Jang [1], Min Su Kim [1], Chang-Lin Li [2] and Tae Hee Han [3],*

[1] Department of Semiconductor and Display Engineering, Sungkyunkwan University, Suwon 16419, Gyeonggi-Do, Korea; sodehd11@skku.edu (J.Y.J.); runemory@skku.edu (M.S.K.)
[2] Eyenix Co., Suwon 16690, Gyeonggi-Do, Korea; clli@eyenix.com
[3] Department of Electrical and Computer Engineering, Sungkyunkwan University, Suwon 16419, Gyeonggi-Do, Korea
* Correspondence: than@skku.edu; Tel.: +82-31-299-4587

Received: 3 March 2019; Accepted: 22 March 2019; Published: 25 March 2019

Abstract: To address the performance bottleneck in metal-based interconnects, hybrid optical network-on-chip (HONoC) has emerged as a new alternative. However, as the size of the HONoC grows, insertion loss and crosstalk noise increase, leading to excessive laser source output power and performance degradation. Therefore, we propose a low-power scalable HONoC architecture by incorporating semiconductor optical amplifiers (SOAs). An SOA placement algorithm is developed considering insertion loss and crosstalk noise. Furthermore, we establish a worst-case crosstalk noise model of SOA-enabled HONoC and induce optimized SOA gains with respect to power consumption and performance, respectively. Extensive simulations for worst-case signal-to-noise ratio (SNR) and power consumption are conducted under various traffic patterns and different network sizes. Simulation results show that the proposed SOA-enabled HONoC architecture and the associated algorithm help sustain the performance as network size increases without additional laser source power.

Keywords: hybrid optical network-on-chip (HONoC); insertion loss; crosstalk noise; signal-to-noise ratio (SNR); semiconductor optical amplifier (SOA)

---

1. Introduction

As multicore systems-on-chips (SoCs) have become mainstream as chips suffer from diminishing returns of miniaturization and the power wall, a new high-performance on-chip network architecture is needed to overcome the physical limitations of metal interconnects [1,2]. A hybrid optical network-on-chip (HONoC) in which the electrical and optical layers are combined, using silicon photonics technology, is emerging as a new alternative for replacing metallic interconnects [3]. The HONoC uses the optical layer to transmit massive data and uses the electrical layer to send control packets or small-size data. As the number of cores in the system increases, the network-size-scalable architecture is highly desirable. The widely-known mesh topology is chosen due to its simplicity and regularity in terms of layout, after considering the advantages of the topology revealed by existing studies on mesh-based HONoCs [4–6].

For reliable communication, the optical signal power arriving at the receiver should be higher than the sensitivity of the photodetector. Because of the significant insertion loss in large-scale HONoCs due to the accumulated microring resonator (MR) drops and waveguide crossings, the laser source output power should be increased. As the laser source is the most power-consuming device among the optical devices, the insertion loss acts as a dominant factor for the overall power consumption of the HONoC.

Lan et al. [7] suggested the dynamic laser power control to reduce the power consumption of the ONoC. However, adaptive laser power management demands complicated control and exhibits limited power efficiency. Moreover, even if the power efficiency of the chip is increased by reducing the laser source power, reliable communication is not guaranteed under high crosstalk noise environments with heavy data traffic cases [8].

We propose a low-power scalable HONoC architecture by deploying semiconductor optical amplifiers (SOAs) and the closely related SOA placement algorithm considering the insertion loss and crosstalk noise change caused by SOA deployment. An analytic worst-case crosstalk noise model is developed to find the optimized SOA gain that minimizes crosstalk noise and to investigate the effect of SOAs on worst-case signal-to-noise ratio (SNR) in HONoC. Considering the relationship between SOA placement and worst-case insertion loss, we calculate the minimum SOA gain for laser source power savings.

The remainder of this paper is organized as follows. Section 2 briefly provides related research and background information regarding insertion loss and crosstalk noise. In Section 3, we introduce the optical router and the SOA model and present a constraint optimization problem for implementing an SOA-enabled HONoC. In Section 4, to solve the constraint optimization problem, we propose an SOA placement algorithm and analyse the worst-case crosstalk noise. Based on the crosstalk noise model, the correlation between the SOA and the SNR is analysed. We also calculate the minimum SOA gain to apply laser source power independent of the longest path. In Section 5, we evaluate and compare the worst-case SNR and power consumption of SOA-enabled HONoC. We draw the conclusion in Section 6.

2. Related Work

2.1. Insertion Loss

The worst-case insertion loss is the dominant factor in determining the laser source output power to ensure reliable communication in the HONoC. The two major causes of insertion loss due to MR drop (Figure 1a) and waveguide crossings (Figure 1b) were reported to be 0.5–1.5 dB and 0.04–0.12 dB, respectively [9–13].

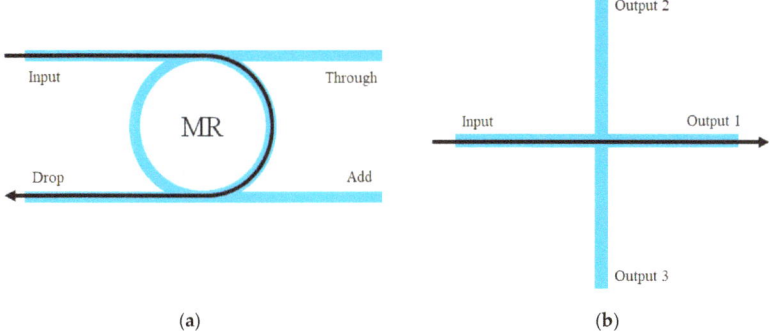

Figure 1. Two major insertion losses in optical network-on-chip (ONoC): (**a**) Microring (MR) drop loss; and (**b**) waveguide crossing loss.

The received signal power should be higher than the sensitivity of the photodetector. Therefore, the relationship between the laser source power (P_{laser}), worst-case insertion loss (P_{worst_IL}) and sensitivity of the photodetector ($P_{sensitivity}$) can be expressed as

$$P_{laser} \geq P_{worst_{IL}} + P_{sensitivity}, \tag{1}$$

where $P_{sensitivity}$ has been reported as approximately -20 dBm [14].

Other types of optical signal power losses considered at the system-level analysis are as follows:

(1) The waveguide propagation loss is -0.274 dB/cm [15].
(2) The MR through loss is smaller than the MR drop loss by two orders of magnitude [16].

Given that these factors are minor in comparison to major insertion losses, most on-chip network architecture studies focus on reducing the insertion losses caused by MR drops and waveguide crossings.

2.2. Crosstalk Noise

Crosstalk noise is an unavoidable characteristic of photonic devices. Sanchis et al. proposed a method for selecting the optimal crossing angle of the waveguide to reduce the crosstalk [17]. Chen et al. proposed multimode-interference (MMI)-based waveguide crossing instead of plain waveguide crossing for reducing signal power loss [18]. They designed a compact structure to minimize the crosstalk at the device level.

Despite these efforts, leakage signals that act as crosstalk cannot be ignored due to their accumulation. Especially, crosstalk noise should be considered as network size increases. In a network-level, Xie et al. analysed crosstalk noise and worst-case SNR through a formal approach in a mesh-based ONoC [19]. According to their research, crosstalk noise causes severe performance degradation and limits the scalability of the HONoC architectures.

Studies on crosstalk noise have been conducted with respect to router architectures, network topologies and traffic patterns [20–22]. On the other hand, there are few studies regarding crosstalk noise when integrating the SOA. The SOA can significantly reduce insertion loss; however, it can also increase crosstalk noise. In this regard, we analyse the correlation between the SOA, insertion loss and crosstalk noise and then propose a methodology to efficiently compensate insertion loss and minimize crosstalk noise.

3. SOA-Enabled HONoC Architecture

Our proposed architecture, as shown in Figure 2, is a mesh-based HONoC with integrated SOAs and consists of two layers. The electrical layer controls the optical layer while transmitting control packets and small-size data through packet switching. Before the optical data transaction between the cores starts, the routing path should be setup in advance through the electrical layer. During the path setup process, the ON/OFF state of SOAs and MRs located in the path are determined and the SOA gain is controlled by adjusting the bias current applied to the SOA.

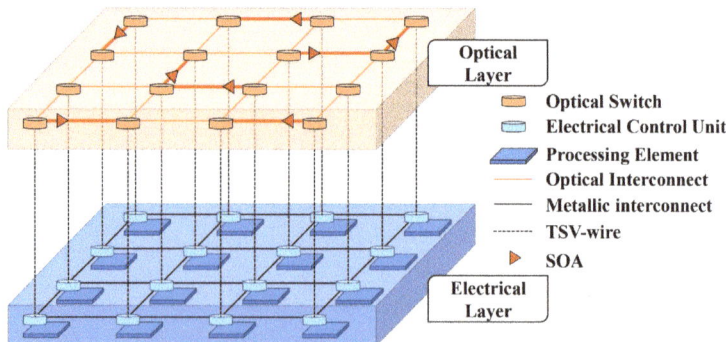

Figure 2. Semiconductor optical amplifier (SOA)-enabled HONoC architecture.

In the optical layer, data transaction is performed using circuit switching due to the absence of optical storage. The SOAs can amplify the optical signal power attenuated by photonic devices in the routing path. The SOAs outside of the routing path are in the OFF state and absorb the optical signal power. In this study, the expected effects of integrating the SOA with the mesh-based HONoC are as follows:

- The SOA compensates for the insertion loss by amplifying the optical signal and reduces the burden of the laser source to reduce the total power consumption of the HONoC.
- An appropriate SOA gain can prevent the SNR degradation problem of the longest path due to the network size growth of the mesh-based HONoC.

3.1. Optical Router Model

The 5 × 5 general optical router for mesh-based HONoC, as shown in Figure 3, consists of the following five ports: Injection/Ejection, North, East, South and West. The ONoC parameters in this analysis are based on the study conducted by Xie et al. [19], which presented the formal analytic models of insertion loss and crosstalk noise for mesh-based HONoCs.

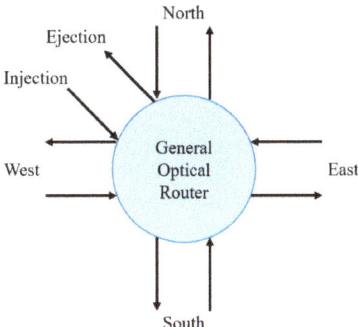

Figure 3. General optical router model.

Optical signals can traverse from the ith port to the jth port and the value of i and j varies from 0 to 4: 0 denotes the Injection/Ejection port and 1–4 denote the North, East, South and West ports, respectively. P_{in}^i denotes the optical power coming into the ith port and $L_{i,j}(x,y)$ is the insertion loss from the ith port to the jth port in the optical router located at (x,y) in the mesh-based HONoC. $P_{i,j}(x,y)$, defined in (2), is the optical power when going from the ith port to the jth port in the optical router $R(x,y)$.

$$P_{i,j}(x,y) = P_{in} L_{i,j}(x,y)\, i,j \in \{0, \ldots, 4\},\ x \in \{1, \ldots, m\},\ y \in \{1, \ldots, n\} \tag{2}$$

$N_{i,j}(x,y)$, calculated in (3), denotes the crosstalk noise introduced into the optical router located at (x,y) in the mesh-based HONoC and $K_{i,j,k}$ is the crosstalk noise introduced into the kth port when the optical signal travels from the ith port to the jth port.

$$N_{i,j}(x,y) = P_{in}^0(x,y)K_{i,j,0} + P_{in}^1(x,y)K_{i,j,1} + P_{in}^2(x,y)K_{i,j,2} + P_{in}^3(x,y)K_{i,j,3} + P_{in}^4(x,y)K_{i,j,4} \tag{3}$$

The SNR is the ratio of the signal power to the noise power and can be written as

$$\text{SNR} = 10\log\left(\frac{P_S}{P_N}\right), \tag{4}$$

where P_S is the power of the optical signal and P_N is the power of noise.

3.2. Semiconductor Optical Amplifier

The SOA is mainly used for amplifying the input optical signals, switching and frequency conversion [23]. The SOA can be integrated into a CMOS-compatible chip and thus can be implemented in the HONoC by using silicon photonics technology. The operating principle of the SOA is that electrons and holes are injected into the n-type region and the p-type region respectively, causing population inversion in the active region. Population inversion leads to stimulated emission and therefore, a photon injected into the SOA triggers another photon with the same phase, wavelength and direction. Optical gain is obtained through this process.

In the active region, the gain per unit length of the SOA is called material gain (g) and is expressed in (5a) as a function of wavelength (λ) and current (I). As L increases, the SOA gain (G) increases accordingly. Here, g and G are in cm^{-1} and dB units, respectively.

$$g(\lambda, I) = [\Gamma a_1 n_0 \left\{ \frac{I}{I_0} - 1 \right\} - \alpha][1 - \frac{2(\lambda - 1570)^2}{\Delta \lambda^2}] \quad (5a)$$

$$G(\lambda, I) = 10 \log_{10}\left(e^{L*g(\lambda,\ I)}\right) \quad (5b)$$

Γ, a_1, n_0, L, I_0, α and $\Delta \lambda$ depend on the material and the structural characteristics of the SOA; the definitions and values of each parameter are shown in Table 1 [24]. As in Equation (5), the SOA gain is proportional to the current. The operating voltage of the SOA and the wavelength are assumed to be 1.5 V and 1550 nm, respectively. In addition, it takes 20–50 ps for the SOA to reach the target gain. The SOA model in this study is simplified for the purpose of system-level analysis; therefore, the physical effect of the amplified spontaneous emission (ASE) introduced by the SOAs is not considered. Although ASE cannot be ignored in physical implementations with contemporary silicon photonics technology, this study focuses on the emerging architecture with a matured manufacturing process that enables feasible on-chip SOAs, assuming a negligible ASE compared to the cumulated crosstalk noise. Additionally, it is also assumed that the placement of SOAs can be over-layered on waveguides in the physical layout; therefore, SOAs do not affect the spaces between optical routers.

Table 1. Definitions and values of SOA parameters.

Parameter	Definition	Value
Γ	Light confinement factor	0.4
a_1	Constant	6.7×10^{-16} cm^2
n_0	Transparency carrier concentration	1.2×10^{18} cm^{-3}
L	Length of the SOA active region	10 µm
I_0	Threshold input current	5 µA
α	Loss in the SOA active region	10 cm^{-1}
$\Delta \lambda$	SOA gain linewidth	95 nm

3.3. Constrained Optimization Problem in SOA-Enabled HONoC

In the HONoC, laser source power mainly depends on the total loss experienced by the optical signal in the longest path. The laser source power level is determined to ensure that the signal power at the receiver is greater than the sensitivity of the photodetector. Therefore, we reduce the burden of the laser source power by using the SOA, which is easy to integrate and of relatively low processing cost. Appropriately placing SOAs across the signal path is an effective way to save the laser source output power in large-scale HONoCs if the additional power consumption of SOAs is reasonably smaller than the reduced amount of the laser source power.

As the total power consumption of the SOA-enabled HONoC depends on the location, spacing and number of SOAs, an efficient SOA placement is needed to achieve the goal. Furthermore, the SOA placement also affects the worst-case SNR because SOAs can amplify the crosstalk noise power as well as the desired optical signal power. Therefore, the constrained optimization problem for implementing the SOA-enabled HONoC in an $m \times n$ mesh is defined as follows:

$$\text{Minimize}: \sum_{total} P_{laser} + \sum_{total} P_{SOA}$$

$$\text{Subject to}: P_{sensitivity} \leq min\left\{PL^{SOA}_{(x_0,y_0)(x_1,y_1)}\right\} \tag{6a}$$

$$min\left\{SNR_{(x_0,y_0)(x_1,y_1)}\right\} \leq min\left\{SNR^{SOA}_{(x_0,y_0)(x_1,y_1)}\right\} \tag{6b}$$

$$x_0, x_1 \in \{1, \ldots, m\}, \ y_0, y_1 \in \{1, \ldots, n\}$$

Here, P_{laser} is the laser source power consumption of the SOA-enabled HONoC, P_{SOA} is the total power consumption of the SOAs, $P_{sensitivity}$ is the sensitivity of the photodetector, $PL^{SOA}_{(x_0,y_0)(x_1,y_1)}$ is the power loss of the optical signal traveling from the router (x_0, y_0) to the router (x_1, y_1) in the SOA-enabled HONoC, $SNR_{(x_0,y_0)(x_1,y_1)}$ is the SNR of the path from the router (x_0, y_0) to the router (x_1, y_1) in the conventional mesh-based HONoC and $SNR^{SOA}_{(x_0,y_0)(x_1,y_1)}$ is the SNR of the path from the router (x_0, y_0) to the router (x_1, y_1) in the mesh-based HONoC with SOAs.

4. Design Methodology for SOA-Enabled HONoC

As mentioned in Section 3.3, the insertion loss problem exacerbates as the network size scales up. On the other hand, the SOA placed to compensate for the insertion loss might adversely affect the SNR caused by unintended crosstalk noise amplification. Therefore, SOA placement considering the routing algorithm should be performed in terms of SNR. In this regard, we devise a design methodology of the SOA-enabled HONoC when XY routing is used in the mesh topology. XY routing has been reported to realize the best performance in terms of bandwidth, latency, load balancing and insertion loss in mesh-based HONoC [5].

The proposed methodology for the SOA-enabled HONoC comprises three steps, as shown in Figure 4. In step 1, we propose an SOA placement algorithm considering insertion loss and SNR. In step 2, the worst-case crosstalk noise due to the SOA placement and the SOA gain to be realized to suppress the crosstalk noise amplification are presented. Based on the SNR analysis with segmented regions, we compare the SNR of the longest path according to the SOA gain and analyse the path in which SNR degradation occurs even if the SOA gain is controlled. In step 3, the algorithm for finding the minimum required SOA gain is developed to allocate the laser source output power independent of the network size.

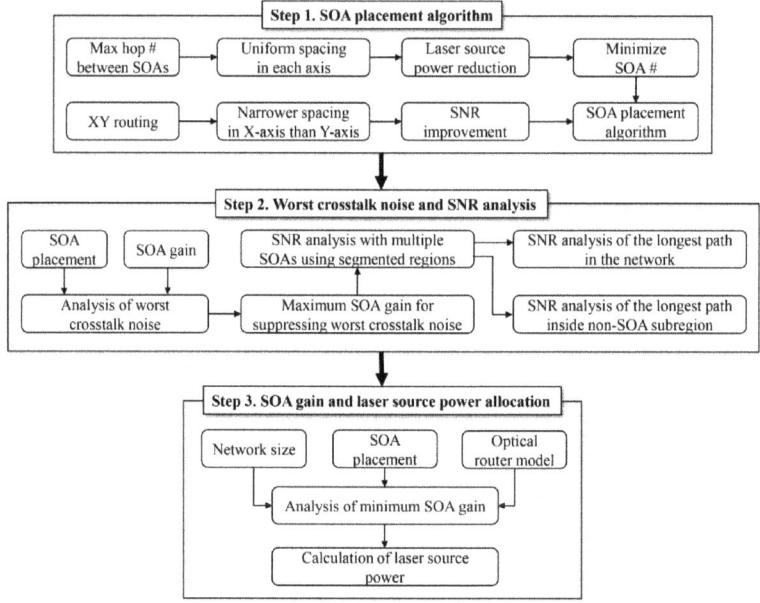

Figure 4. Overall flow of the proposed design methodology.

4.1. SOA Placement Algorithm for Laser Source Power Saving and SNR Enhancement

The SOA gain modelling in Equation (5) shows that using a single SOA is superior to using multiple SOAs in that it uses less bias current to obtain the same gain, as described in Equation (7).

$$G(I_{total}) > \sum_{i=1}^{n} G(I_i), \quad I_{total} = \sum_{i=1}^{n} I_i \qquad (7)$$

In addition to the gain efficiency of the SOA, the small number of SOAs with high gain is advantageous in terms of the area in which the SOA devices and control units are placed. However, this strategy should be validated. As the worst-case insertion loss determines the output power of laser sources and even one SOA has impact on various routing paths, an effective SOA placement rule is needed to reduce the laser source power using the appropriate number of SOAs. Considering the insertion loss due to the SOA placement in the routing path, the power consumption of the HONoC can be effectively reduced by equalizing the maximum number of hops that do not pass through the SOA in all routing paths. We denote this maximum hop count without SOAs as h, which is determined by t_x and t_y, indicating the SOA spacing on the X-axis and Y-axis, respectively.

The accumulated crosstalk noise power at the destination depends on not only the number of SOAs but also the location of the SOAs. For example, as shown in Figure 5, the crosstalk power increases when the SOA is placed nearer to the destination rather than the source. In other words, to minimize the crosstalk power amplified through the SOAs in the optical signal path, SOA should be placed close to the source. In the mesh-based HONoC with XY routing, the optical signal is transmitted along the Y-axis after the X-axis. Hence, placing more SOAs on the X-axis is preferred to Y-axis in terms of SNR. Accordingly, we propose the SOA placement algorithm as follows. In this study, it is assumed that all SOA gains are the same for the convenience of design and analysis of the network.

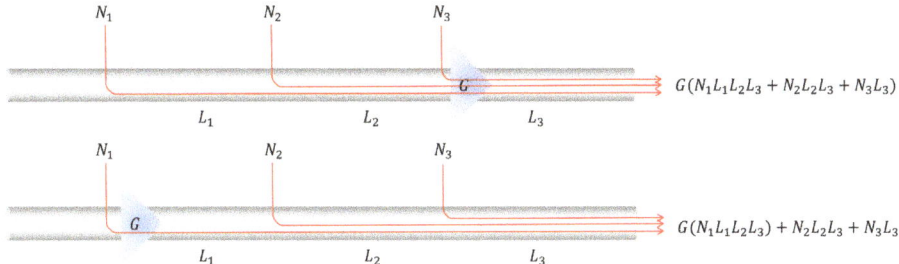

Figure 5. Example of cumulated crosstalk noise according to the SOA placement.

In Algorithm 1, the optimal SOA spacing in the X-axis (t_x) and that in the Y-axis (t_y) are selected according to the maximum number of hops among the possible routing paths without SOAs (h). Therefore, the total number of SOAs placed in the network depends on the possible combinations of t_x and t_y with a consistent h. To minimize the additional power consumption of the SOAs, use of the smallest number of SOAs should be pursued. If there are several cases with the same number of SOAs, the narrowest X-axis spacing is chosen to improve the SNR (lines 1–10). After the spacing of the SOAs is determined, SOAs are placed on the link of the corresponding router (lines 11–23).

Algorithm 1. SOA Placement for Laser Source Power Saving and SNR Improvement

Input: mesh size (m, n), maximum hop count of possible paths without SOAs (h)
Output: number of SOAs (n_{SOA}), SOA spacing of X-axis and Y-axis (t_x, t_y)

1:   $n_{SOA} = (n-1)/1 + (m-1)/(h+1)$
2:   **For** i from 1 to $h+1$ **do**
3:     $j = h + 2 - i$
4:     $n[i] = (n-1)/i + (m-1)/j$
5:     **If** $n[i] < n_{SOA}$ **then**
6:       $n_{SOA} = n[i]$
7:       $t = i$
8:     **End if**
9:   **End for**
10: $t_x = t$, $t_y = h + 2 - t$, $n_{SOA} = m \cdot t_x + n \cdot t_y$
11: **For** i from t_x to $n-1$ **do**
12:    **For** j from 1 to m **do**
13:      Allocate SOA at east link of R(j, i)
14:      $j = j + 1$
15:    **End for**
16:    $i = i + t_x$
17: **End for**
18: **For** i from t_y to $m-1$ **do**
19:    **For** j from 1 to n **do**
20:      Allocate SOA at south link of R(i, j)
21:      $j = j + 1$
22:    **End for**
23:    $i = i + t_y$
24: **End for**

4.2. Worst-Case Crosstalk Noise and SNR Analysis

For the analysis of the worst-case SNR in the proposed architecture, we use several definitions as follows:

- n_r: the total number of routers in the optical signal path

- n_s: the total number of SOAs in the optical signal path
- t_i: the number of the routers between the ith SOA and the $(i-1)$th SOA in the optical signal path ($t_0 = 0$, t_{n_s+1} = (# of routers between the last SOA and the destination))
- P_{N_i}: the crosstalk noise introduced into the ith router in the optical signal path
- L: router loss, G: SOA gain, K: crosstalk noise coefficient

To simplify the worst-case SNR equations, we make the following assumptions

- The insertion loss and the crosstalk noise coefficient are the same regardless of ports. ($L_{i,j} = L, K_{i,j,n} = K$)
- The mesh size is $m \times m$.

As the crosstalk noise introduced into each router is also amplified by the corresponding SOAs in the optical signal path, these effects are reflected in the SNR as in (8).

$$SNR = 10\log\left(\frac{P_S G^{n_s} L^{n_r}}{\sum_{i=0}^{n_s} G^{n_s-i} \sum_{j=T_i+1}^{T_{i+1}} \left(P_{N_j} L^{n_r-j}\right)}\right) \text{ where } T_i = \sum_{k=0}^{i} t_k \quad (8)$$

As shown in Equation (8), it is inevitable that the crosstalk noise introduced into the optical signal passes through the SOAs located in the signal path. Therefore, to maximize the SNR, the worst crosstalk candidate must be suppressed before it leaks into the desired optical signal path.

The dominant crosstalk power is mainly caused by the crosstalk noise coming from the nearest router in conventional HONoCs. However, in the SOA-integrated HONoC, the worst crosstalk candidates vary depending on the gain of the associated SOAs. If the SOA gain is larger than the insertion loss between the SOAs, the crosstalk noise introduced from a farther router has a greater impact on the SNR than that from a nearer router. For example, when there are the three crosstalk noise sources P_1, P_2 and P_3 as shown in Figure 6, each crosstalk noise power can be expressed as in (9). If the total SOA gain is smaller than the total power loss ($\prod G_i L_i < 1$), the magnitude of each crosstalk noise power is ordered as $N_1 < N_2 < N_3$; otherwise, $N_1 > N_2 > N_3$.

$$N_1 = P_1 G_1 L_1 G_2 L_2 G_3 L_3 K \quad (9a)$$

$$N_2 = P_2 G_2 L_2 G_3 L_3 K \quad (9b)$$

$$N_3 = P_3 G_3 L_3 K \quad (9c)$$

where $P_1 = P_2 = P_3, G_1 = G_2 = G_3$

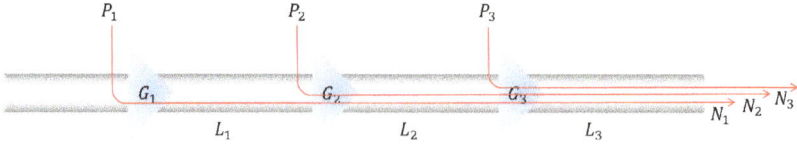

Figure 6. Example of crosstalk noise paths whose magnitude order varies depending on SOA gains.

Therefore, the worst crosstalk noise differs depending on the SOA gain even with the same placement of SOAs. To quantify the effect of SOAs when the current signal path acts as a crosstalk noise to other adjacent signal paths, we introduce the concept of average insertion loss (L_{avg}) per SOA as (10). Here, definitions of terms are as follows:

- $n_{r,c}$: the total number of routers in the crosstalk path
- $n_{s,c}$: the total number of SOAs in the crosstalk path

- $t_{i,c}$: the number of routers between the ith SOA and the $(i-1)$th SOA in the crosstalk path

$$L_{avg} = \begin{cases} \sqrt[n_{s,c}]{\prod_{i \in \{2,\cdots,n_{r,c}\}} L_i}, & \sum_{i=1}^{n_{s,c}} t_{i,c} \neq n_{r,c} \\ \sqrt[n_{s,c}-1]{\prod_{i \in \{2,\cdots,n_{r,c}\}} L_i}, & \sum_{i=1}^{n_{s,c}} t_{i,c} = n_{r,c} \end{cases} \quad (10)$$

To analyse the worst crosstalk candidates, we denote the router to be analysed as R_{DUT} and the SOA farthest from the R_{DUT} in the crosstalk-inducing path as SOA_{DUT}. In Equation (10), L_{avg} is determined by the power loss of an optical signal traveling from SOA_{DUT} to R_{DUT} and the number of SOAs in that path. If an SOA is placed in the link directly connected to R_{DUT}, this SOA is not included in the calculation for L_{avg} because incoming crosstalk noise passes through that SOA inescapably. With the SOA gain and the average insertion loss, the dominant crosstalk noise introduced into the router can be identified. The worst crosstalk noise due to the SOA gain can be expressed as Equation (11).

(1) $G > |L_{avg}|$,
$$P_N = P_{Laser} K L^{n_{r,c}} G^{n_{s,c}} \quad (11a)$$

(2) $G \leq |L_{avg}|$,
$$P_N = \begin{cases} P_{Laser} KL, & \sum_{i=1}^{n_{s,c}} t_{i,c} \neq n_{r,c} \\ P_{Laser} KLG, & \sum_{i=1}^{n_{s,c}} t_{i,c} = n_{r,c} \end{cases} \quad (11b)$$

If the SOA gain is greater than $|L_{avg}|$, the crosstalk power originating from the farthest SOA is the largest. This worst crosstalk noise power increases as the SOA gain and mesh size increases and the SOA spacing decreases. Conversely, if the SOA gain is less than or equal to $|L_{avg}|$, the crosstalk noise from the nearest router is dominant. In this case, the crosstalk can be amplified or not, depending on the presence of the SOA in the corresponding waveguide.

Therefore, SOA gain control is required to minimize the dominant crosstalk before it leaks into the optical signal path. The average insertion loss is equal to the maximum gain of the SOA required to minimize the worst crosstalk as given in Equation (12).

$$G_{max} = |L_{avg}| \quad (12)$$

On the other hand, as the crosstalk noise introduced into the optical signal is amplified by the SOAs on the signal path, we analyse the specific cases in which SNRs are strongly affected by the SOAs.

4.2.1. SNR Analysis with Multiple SOAs Using Segmented Regions in Mesh-Based HONoCs

To facilitate the analysis of the SNR change resulting from SOAs and crosstalk, the entire on-chip optical network is partitioned into several sub regions. The criteria for region partitioning are that no sub region should contain any SOA inside and should be surrounded by the SOA-placed links. The sub region is denoted as $REG(i,j)$ where i and j indicate the row and column position, respectively, as shown in Figure 7.

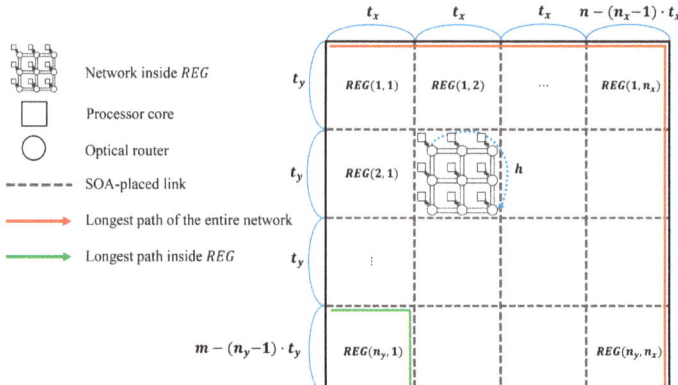

Figure 7. Non-SOA sub regions (*REG*) according to the SOA placement algorithm.

It is apparent that the crosstalk noise introduced at the same *REG* with the source node of the optical signal to be routed will be subject to the same number of SOAs as the optical signal, whereas the other crosstalk noises introduced at nearby *REGs* to the destination node will be amplified to a lesser degree by the SOAs than by the optical signal under analysis. More specifically, we formulated the SNR of the optical signal path that travels from $REG(1,1)$ to $REG(n_x, n_y)$ in Equation (13a) for an SOA-enabled HONoC. For comparison purposes, the SNR for the non-SOA case is derived in Equation (13b).

$$SNR_1 = 10\log\left(\frac{P_S L_{REG(1,1)(n_x,n_y)}}{\sum_{\forall i}\frac{N_{REG(1,i)}L_{REG(1,i+1)(n_x,n_y)}}{G^{(i-1)}} + \sum_{\forall j}\frac{N_{REG(j,n_x)}L_{REG(j+1,n_x)(n_x,n_y)}}{G^{(n_x-1)+(j-1)}}}\right) \quad (13a)$$

$$SNR_2 = 10\log\left(\frac{P_S L_{REG(1,1)(n_x,n_y)}}{\sum_{\forall i} N_{REG(1,i)}L_{REG(1,i+1)(n_x,n_y)} + \sum_{\forall j} N_{REG(j,n_x)}L_{REG(j+1,n_x)(n_x,n_y)}}\right) \quad (13b)$$

$$i \in \{1,\ldots,n_x-1\}, \quad j \in \{1,\ldots,n_y\}$$

Here, the total crosstalk noise introduced into the optical signal when it passes through $REG(i,j)$ is represented by $N_{REG(i,j)}$ and the signal power loss that occurs when migrating from $REG(i_0,j_0)$ to $REG(i_1,j_1)$ is denoted by $L_{REG(i_0,j_0)(i_1,j_1)}$. Additionally, n_x and n_y are the number of subregions divided by the SOA-placed links in the *X* and *Y* axes, respectively.

In the non-SOA enabled HONoC, the SNR decreases as the routing path length increases as a result of increased signal attenuation and more involved crosstalk noises. In the proposed architecture, the effect of additional crosstalk noise according to the path length increase decreases by the appropriate placement of SOAs to boost the SNR. As shown in Equation (13a), the noise terms decrease by G^0, $G^1,\ldots$ and $G^{n_x+n_y-2}$, respectively. The SNR enhancement becomes more apparent as the number of *REGs* increases. Therefore, at the router level, we analysed a path that passes through the greatest number of *REGs* and a path inside an *REG*.

4.2.2. SNR Analysis of the Longest Path

As the worst-case SNR of the longest path can vary depending on the dominant crosstalk affected by the SOA gain, the SNR values of three representative cases are analysed as follows.

(1) Case 1: $h = 0$, $G \leq |L_{avg}|$

$$SNR_1 \approx \frac{L/K}{\left[\left((GL)^{-1} + 2GL\right)\left(\sum_{i=0}^{2m-3} \frac{1}{(GL)^i}\right) - (GL)^{-(m-2)} + (GL)^{-(m-4)}\right]} \quad (14a)$$

(2) Case 2: $h = 0$, $G > |L_{avg}|$

$$SNR_1 \approx \frac{L/K}{\left[\left((GL)^{-1} + (GL)^{m-1}\right)\left(\sum_{i=0}^{2m-3} \frac{1}{(GL)^i}\right) + \left((GL)^{m-1} + 1\right)\left(\sum_{i=0}^{m-2} \frac{1}{(GL)^{2i}}\right)\right]} \quad (14b)$$

(3) Case 3: $h = 2m - 1$

$$SNR_3 \approx \frac{L/K}{\left[(L^{-1} + 2L)\left(\sum_{i=0}^{2m-3} \frac{1}{L^i}\right) - L^{-(m-2)} + L^{-(m-4)}\right]} \quad (14c)$$

The case when $h = 0$ indicates that SOAs are placed on all waveguides, whereas there is no SOA when $h = 2m - 1$. When $h = 0$, a *REG* consists of only one router. As shown in Figure 8a of Case 1, the incoming crosstalk from the nearest router is dominant. In contrast, the crosstalk introduced from the farthest router causes the worst-case SNR in Case 2 as shown in Figure 8b.

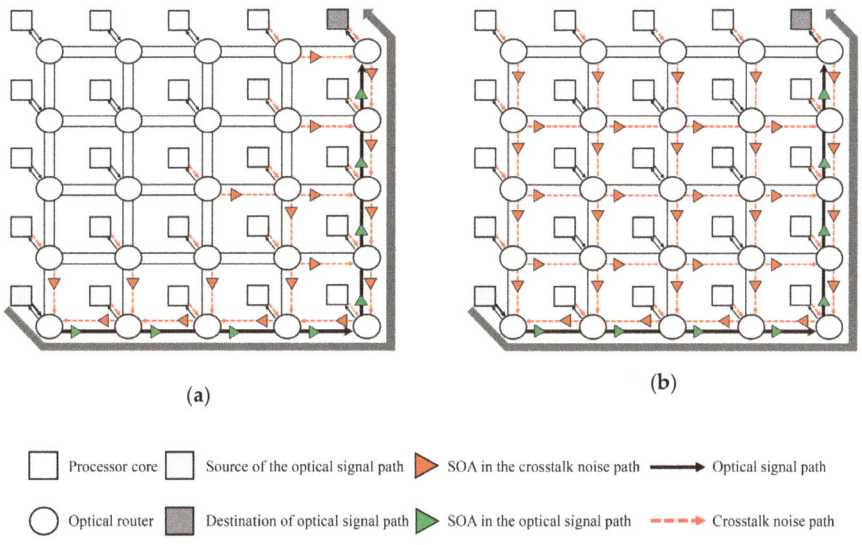

Figure 8. Worst-case SNR of the longest path in accordance with the SOA gain: (**a**) $h = 0$, $G \leq |L_{avg}|$; and (**b**) $h = 0$, $G > |L_{avg}|$.

As shown in Equation (14a) and (14c), SNR_1 and SNR_3 have an identical form because of the same traffic pattern for the worst-case SNR in the network. If the GL of SNR_1 is replaced by L_1 and if the L of SNR_3 is L_3, then L_1 is always less than L_3. Therefore, SNR_1 is always greater than SNR_3 because the insertion loss of Case 1 is smaller than that of Case 3.

In contrast, some crosstalk components have the chance to be amplified more times than the optical signal by SOAs as described in Case 2. In Equation (14b), $(GL)^{m-1}$ represents the worst crosstalk noise due to the SOAs and $(GL)^i$ indicates the ratio of the number of SOAs that the optical signal passes through to the number of the SOAs that the introduced crosstalk noise passes through. Therefore, if $(GL)^{m-1}$ is larger than $(GL)^i$, this implies that the number of SOAs that the crosstalk noise passes through is greater than the number of SOAs that the optical signal passes through. Consequently, SNR_2 severely degrades and eventually can be smaller than SNR_3 as the SOA gain grows.

4.2.3. SNR Analysis inside Regions without SOAs

As previously described, the SNR can be improved when the optical signal passes through multiple SOAs with suitable SOA gain control. However, if the optical signal path forms within a certain REG, the SNR can degrade compared to a non-SOA HONoC, even if the condition $G < |L_{avg}|$ is satisfied.

When there is a path that travels along the boundary of REG, as shown in Figure 9, both the crosstalk noise occurring inside REG and the optical signal will not pass through any SOA, whereas the crosstalk noise originating from the outside of REG is amplified by SOAs.

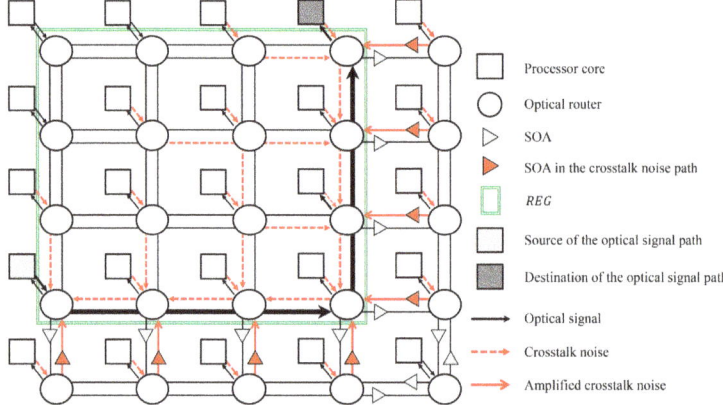

Figure 9. Worst-case signal to noise ratio (SNR) of the non-SOA region surrounded by SOA-placed links.

The worst-case SNR of the SOA-integrated HONoC under the corresponding traffic condition and that of the HONoC without SOAs under the same condition can be expressed as follows:

(1) Case 1: $0 < h < 2m - 1$, $G < |L_{avg}|$

$$SNR_1 \approx \frac{L^{t_x+t_y-1}/K}{\left[\begin{array}{c} GL\left(2\sum_{i=1}^{t_x} L^{t_x+t_y-1-i} + \sum_{i=2}^{t_y} L^{t_y-i}\right) + 2\left(\sum_{i=1}^{t_x-2} L^{t_x+t_y-i} + \sum_{i=1}^{t_y-2} L^{t_y-i}\right) \\ + \left(L^{t_y+1} + L^{t_y} + L\right) + \left(\sum_{i=1}^{t_x+t_y-2} L^{t_x+t_y-2-i}\right) \end{array}\right]} \quad (15a)$$

(2) Case 2: $h = 2m - 1$

$$SNR_2 \approx \frac{L^{t_x+t_y-1}/K}{\left[\begin{array}{c} L\left(2\sum_{i=1}^{t_x} L^{t_x+t_y-1-i} + \sum_{i=2}^{t_y} L^{t_y-i}\right) + 2\left(\sum_{i=1}^{t_x-2} L^{t_x+t_y-i} + \sum_{i=1}^{t_y-2} L^{t_y-i}\right) \\ + \left(L^{t_y+1} + L^{t_y} + L\right) + \left(\sum_{i=1}^{t_x+t_y-2} L^{t_x+t_y-2-i}\right) \end{array}\right]} \quad (15b)$$

Case 1 represents the SOA-integrated HONoC and Case 2 denotes the HONoC without SOAs. In Equation (15a), the first term in the denominator is the crosstalk noise introduced from outside of REG and amplified by the SOA gain (G). Assuming the same optical signal power for both cases, the SNR of the SOA-integrated HONoC is less than that of the HONoC without SOAs. Therefore, if the SOAs are not placed on every link, there is a path along which the SNR decreases.

4.3. Minimum SOA Gain and Laser Source Power Allocation

In a conventional mesh-based HONoC, a laser source power is applied considering the insertion loss of the longest path and the receiver sensitivity. However, in the proposed SOA-enabled HONoC, a new analysis of the worst-case insertion loss is required, as the optical signal is amplified whenever it passes through the SOA. If the total SOA gain is larger than the total insertion loss of the overall signal path, the signal power at the destination is greater than that at the source. Therefore, we present the allocation of the laser source power independent of the longest path considering the SOA placement, SOA gain and mesh size.

4.3.1. Minimum SOA Gain Allocation

The SOA gain should be sufficiently large to compensate for the insertion loss between two adjacent SOAs along the signal path. Thus, the laser source power need not be increased even if the network size grows. Overall, the laser source power level is determined by the insertion loss of the longest signal path that does not pass through any SOAs.

As shown in Figure 10, the top-leftmost sub region $REG(1,1)$ occupies the largest area among the sub regions because uniform placement of SOAs is performed from the top-leftmost corner of the network according to the proposed SOA placement algorithm. Assuming that the reference laser source power level is calculated by considering the longest path formed within $REG(1,1)$, we can compute the minimum SOA gain to compensate for the insertion loss without increasing the laser source power.

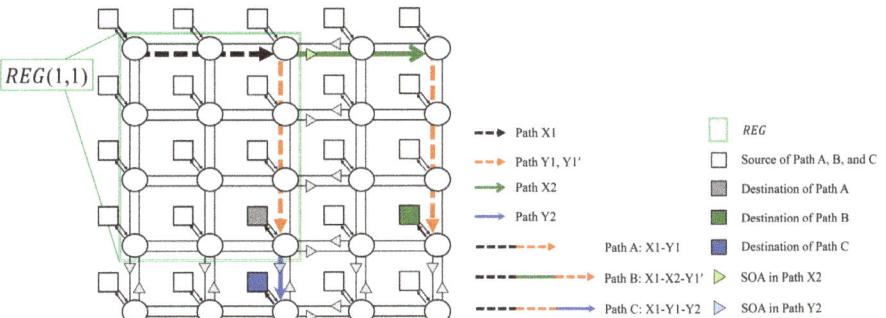

Figure 10. Example of determining the minimum SOA gain considering traffic paths across the top-leftmost non-SOA region.

Once the laser source power is determined by the insertion loss of Path A, the next step is to identify the larger insertion loss between Path B and Path C to compute the minimum SOA gain. Finally, the minimum SOA gain is chosen as the larger value between the insertion loss of Path X2 and that of Path Y2. To summarize, the minimum SOA gain can be expressed as follows:

$$G_{min} = max(L_{X2}, L_{Y2}), \qquad (16)$$

where L_{X2} and L_{Y2} denote the insertion loss of Paths X2 and Y2, respectively. Considering mesh size and SOA spacing, we propose the following algorithm to calculate the minimum SOA gain:

As the optical signal passes through the router in a straight manner, Algorithm 2 checks the paths (W → E) and (E → W) in the X-axis and (S → N) and (N → S) in the Y-axis, respectively, to find the maximum power loss. Then, L_{X2} and L_{Y2} are independently calculated for each axis. If $n_x \geq 2$, the maximum hop count is the SOA spacing along the X-axis (t_x) (lines 2–3). Otherwise, the hop count is the number of routers in the X-axis (n) excluding t_x (lines 4–6). The same process is performed on the Y-axis (lines 7–11) and the larger insertion loss between L_{X2} and L_{Y2} is chosen as the minimum SOA gain (line 12).

Algorithm 2. Calculation for the Minimum SOA Gain

Input: mesh size (m, n), SOA spacing (t_x, t_y)
Output: minimum SOA gain (G_{min})
1:     $n_x = (n-1)/t_x$, $n_y = (m-1)/t_y$
2:     If $n_x \geq 2$ then
3:         $L_{X2} = max(L_{W,E}, L_{E,W}) \times t_x$
4:     Else
5:         $L_{X2} = max(L_{W,E}, L_{E,W}) \times (n - t_x)$
6:     End if
7:     If $n_y \geq 2$ then
8:         $L_{Y2} = max(L_{S,N}, L_{N,S}) \times t_y$
9:     Else
10:    $L_{Y2} = max(L_{S,N}, L_{N,S}) \times (m - t_y)$
11:   End if
12:   $G_{min} = max(L_{X2}, L_{Y2})$

4.3.2. Minimum Laser Source Power Allocation

When the SOA effectively compensates for the insertion loss along the optical signal path, the worst-case insertion loss depends on h, which is the maximum number of hops along which the optical signal does not pass through any SOAs. If $h = 0$, the maximum insertion loss case occurs when the length of the optical signal path is one hop as the optical signal is amplified just once. Otherwise, the worst insertion loss is among the sub-longest paths that are not amplified by the SOA.

When the waveguide crossing loss is –0.12 dB, the insertion loss of the input/output port in the 5 × 5 Crux router is shown in Table 2. The crux router is the optimized optical router for the insertion loss and the SNR when using X-Y routing [25]. A shown Table 2, the optical router has different insertion loss values depending on the signal flow at the input/output ports. Therefore, the total power loss depends on the routing path even if the number of hops that the optical signal travels along are the same. The worst-case insertion loss according to h considering the input and output port is as follows.

(1) $h = 0, 1$

$$P_{worst\_IL} = (L_{In,E} + L_{W,Ej})\ or\ (L_{In,N} + L_{S,Ej}) \tag{17a}$$

(2) $h = 2$

$$P_{worst\_IL} = (L_{In,E} + L_{W,N} + L_{S,Ej}) \tag{17b}$$

(3) $h \geq 3$

$$P_{worst\_IL} = (L_{In,E} + L_{W,N} + L_{S,N} \times (h-2) + L_{S,Ej}) \tag{17c}$$

Here, the loss values are in dB units. When $h = 0$, the worst-case insertion loss is equal to the case of $h = 1$, as the SOA exists in all links; however, the SOA gain should be additionally considered when calculating the minimum laser source power. Therefore, the minimum laser source power according to h can be expressed as follows:

$$P_{laser\_min} = \begin{cases} P_{worst\_IL} + P_{sensitivity} - P_{gain}, & h = 0 \\ P_{worst\_IL} + P_{sensitivity}, & h \neq 0 \end{cases} \tag{18}$$

where P_{worst_IL} is the worst-case insertion loss in the SOA-enabled HONoC, $P_{sensitivity}$ is the minimum optical power required by the photodetector and P_{gain} is the SOA gain.

Table 2. Loss value in 5 × 5 Crux router.

Loss	Value (dB)	Loss	Value (dB)
$L_{In,W}$	0.50	$L_{E,W}$	0.38
$L_{In,E}$	0.88	$L_{E,N}$	0.50
$L_{In,N}$	0.88	$L_{E,S}$	1.00
$L_{In,S}$	0.63	$L_{E,Ej}$	0.63
$L_{W,E}$	0.38	$L_{N,S}$	0.38
$L_{W,N}$	1.00	$L_{N,Ej}$	0.50
$L_{W,S}$	0.50	$L_{S,N}$	0.38
$L_{W,Ej}$	0.88	$L_{S,Ej}$	0.88

5. Simulation Results

We built up a C-based simulator to evaluate the SNR and the power consumption of the SOA-enabled mesh-based HONoC in comparison with an HONoC without SOAs. The typical values of insertion loss and crosstalk coefficient for system analysis are presented in Table 3. The chip size of 16 × 16 HONoC is assumed as 1 cm^2 and in other cases, the chip sizes are assumed to be proportional to the number of cores. It is reasonable to suppose that the chip size of the SOA-enabled HONoC is the same as that of the conventional one, as the SOA is small enough to be embedded in a waveguide without widening the space between photonic devices. The traffic patterns are generated using modified BookSim [26] and Crux router is deployed because it is optimized for insertion loss and SNR in the mesh topology.

Table 3. Loss and crosstalk coefficient values.

Parameter	Value	Reference
Waveguide crossing loss	−0.12 dB	[12]
Propagation loss per cm	−0.274 dB/cm	[15]
Power loss of CSE in OFF state	−0.04 dB	[13]
Power loss of CSE in ON state	−0.5 dB	[9]
Power loss of PSE in OFF state	−0.005 dB	[9]
Power loss of PSE in ON state	−0.5 dB	[9]
Crossing crosstalk coefficient	−45 dB	[27]
Crosstalk coefficient of PSE in OFF state	−20 dB	[28]
Crosstalk coefficient of PSE in ON state	−25 dB	[28]

5.1. SNR Analysis

In the conventional m × n mesh-based HONoC, it is proved formally that the worst-case SNR is among the 1st, 2nd and 3rd longest paths and the best SNR performance is achieved when m = n [19]. For fair comparison, symmetric mesh topologies are chosen for evaluations.

Figure 11 shows worst-case SNR for the 1st, 2nd and 3rd longest paths according to the SOA gain when every link contains an SOA and the network size is 8 × 8, 16 × 16, 24 × 24 and 32 × 32, respectively. When the SOA gain is 0 dB, the SNR values are the same as that for non-SOA HONoCs. The SNR gradually improves as the SOA gain increases, whereas it decreases at a certain threshold of the SOA gain. Furthermore, the SNR decreases further than that of the conventional HONoC without SOAs if the SOA gain increases even more.

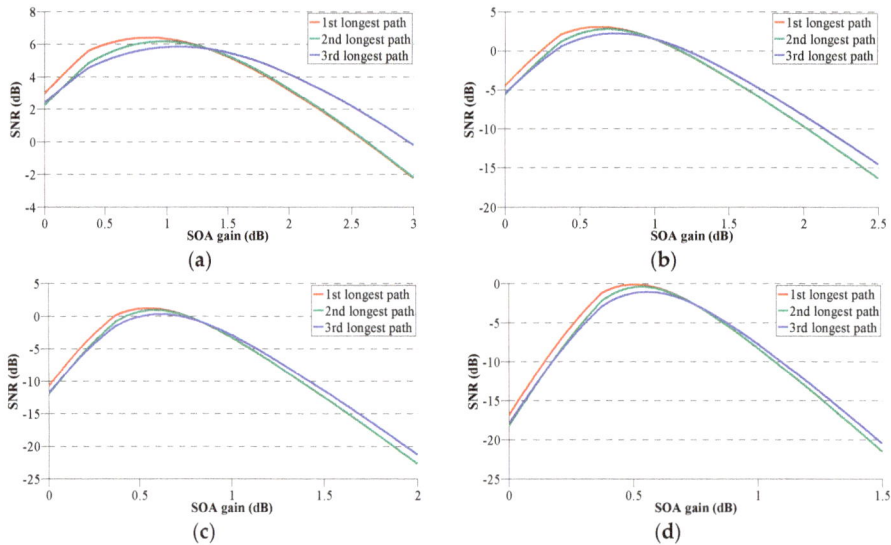

Figure 11. Worst-case SNR for SOA-enabled HONoC according to the SOA gain for different network sizes: (**a**) 8 × 8; (**b**) 16 × 16; (**c**) 24 × 24; and (**d**) 32 × 32.

Figure 12 shows the worst-case SNR of the SOA-enabled HONoC and that of the conventional HONoC according to the mesh size in an m × m mesh. When the network size is larger than 12 × 12, the noise power exceeds the signal power in the mesh-based HONoC. In addition, the SNR decreases drastically as the network size increases. However, in our proposed architecture, the SNR of the 1st, 2nd and 3rd longest paths are 1.19 dB, 0.89 dB and 0.28 dB when the network size is 24 × 24. When the network size is 32 × 32, the worst-case SNR is −1.05 dB and the SNR degradation is not large even if the network size is further increased. As the worst-case SNR determines the feasibility and the scalability of the HONoC, the proposed architecture shows a great advantage over the conventional HONoC architecture.

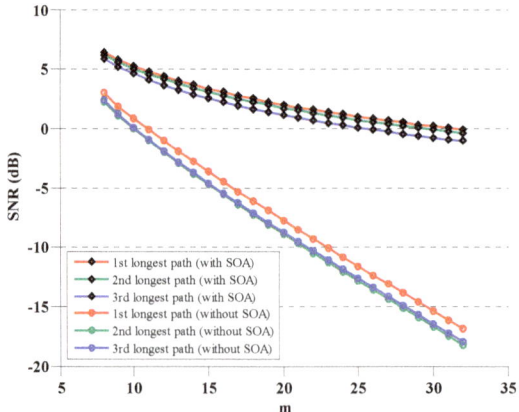

Figure 12. SNR comparison for the 1st, 2nd and 3rd longest paths between SOA-enabled HONoC and conventional HONoC (without SOAs) in accordance with the mesh size.

Figure 13 shows the SNR improvement in an average basis for the 1st, 2nd and 3rd longest paths with respect to the maximum hop count without SOAs, h. The number of SOAs is proportional to the mesh size and is inversely proportional to h. When the mesh size is smaller and h is larger, the amplification of the optical signal is relatively small. On the contrary, the amplification of the introduced crosstalk noise occurs frequently when the mesh size is larger and h is smaller. Therefore, the SNR improvement is closely related to the mesh size as well as h, which determines the number of SOAs in the optical signal path. When the SOA placement is performed according to the optimal h, the SNR is improved by 4.26 dB, 8.91 dB, 13.69 dB and 18.65 dB when the network size is 8×8, 16×16, 24×24 and 32×32, respectively. Therefore, the degradation of the worst-case SNR in the large-scale HONoC can be mitigated through the placement of SOAs.

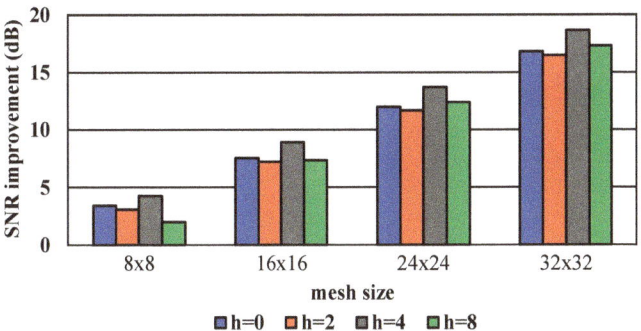

Figure 13. SNR improvement in an average basis for the 1st, 2nd and 3rd longest paths of the SOA-enabled HONoC according to different h and network sizes.

Figure 14 shows the worst-case SNR reduction of the non-SOA region surrounded by the SOA-placed links, in accordance with the SOA gain. As previously analysed in Section 4.2.2, there is a path in which the SNR decreases due to the several amplified crosstalk noises, compared to the equivalent optical signal power if the SOAs are not placed on all links ($h \neq 0$).

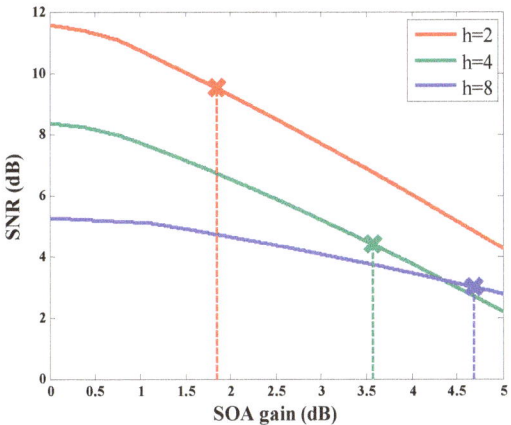

Figure 14. SNR reduction in the region surrounded by the SOA-placed links. X indicates the appropriate SOA gain for the worst-case SNR.

When h becomes smaller, the SNR of the longest path traveling along the boundary decreases more as the SOA gain increases. This result denotes that the shorter path length is more affected by the amplified crosstalk noise. However, as marked X in Figure 14, the appropriate SOA gain for the worst-case SNR on the whole network differs depending on h. The SNR losses at the mark X are 2.06 dB, 3.96 dB and 2.26 dB when $h = 2$, $h = 4$ and $h = 8$, respectively. The SNR reduction is the smallest when $h = 2$; however, there are more paths in which such SNR degradation occurs in the network because the SOA spacing is narrower than in other cases. In other words, it means that SNR degradation may occur when SOA spacing is narrow but it does not occur in the same path when SOA spacing is wide. Therefore, $h = 8$ can be the best case considering both the magnitude of SNR reduction and the probability that these SNR-reduced paths will occur.

5.2. Power Consumption

Figure 15 shows the laser source power considering the mesh size, the placement of the SOA and the worst-case insertion loss. We assume that the efficiency of the laser source power is 20% [29]. If there is no SOA, the laser source power increases sharply as the mesh size increases. In the SOA-enabled HONoC, however, the laser source power is independent of the mesh size. Although additional power consumption due to the SOAs will be inevitable, it is expected to be a comparable trade-off owing to the laser source power savings that will be realized by incorporating this work.

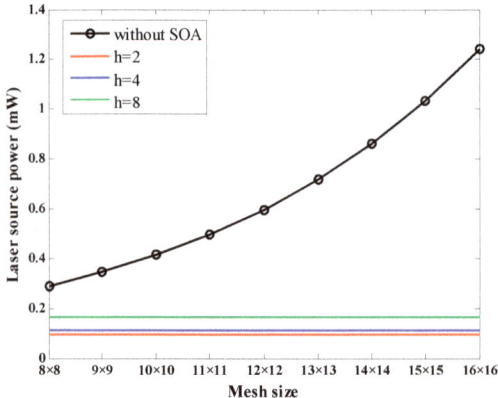

Figure 15. Laser source power comparison between a conventional HONoC (without SOA) and SOA-enabled HONoC.

With common traffic patterns for NoCs, the simulation results in terms of power consumption in the conventional HONoC and the SOA-enabled HONoC are illustrated in Figure 16. We assume that the mesh size is 8×8 and SOAs are placed on every waveguide where the laser source power is the smallest. According to the previous studies, the power consumption of MR activation has been reported as 50 μW [30]. Simulation was performed on the HONoC without SOAs and the SOA-enabled HONoCs optimized for SNR and power consumption, respectively. Power consumption is reduced by 16.26–35.86% when focusing on the SNR and by 21.81–37.68% when focusing on the total power consumption. There is a slight difference in power consumption between when optimizing for SNR and when optimizing for power consumption.

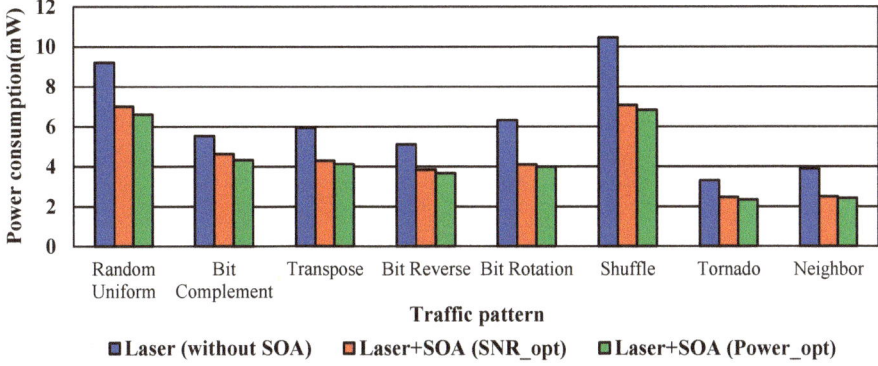

Figure 16. Power consumption comparison between a conventional HONoC (without SOA) and SOA-enabled HONoC optimized for SNR and power consumption, respectively, for various traffic patterns.

Figure 17 shows the total power consumption according to the maximum hop count of possible paths without SOAs (h) and different traffic patterns when network size is 8×8. When $h = 0$, the laser source power can be reduced by the SOA gain; however, the efficiency of the SOA gain is low and the number of SOAs is the most. If $h \neq 0$, the number of SOAs in the network is reduced as h increases; however, the laser source power and single SOA power consumption increase.

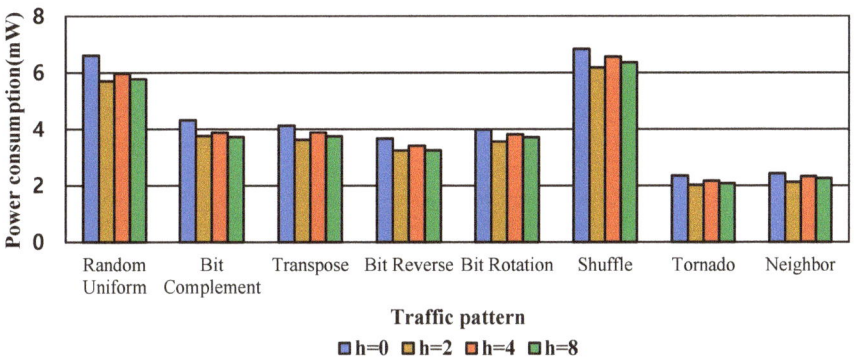

Figure 17. Power consumption of an SOA-enabled HONoC according to the SOA spacing for various traffic patterns.

When $h = 4$, the number of SOAs is twice as high as when $h = 8$ and the power consumption increases. Furthermore, in the case of $h = 2$, the number of SOAs is three times higher than when $h = 8$; however, the power consumption is reduced slightly or similar. Therefore, the total power consumption in relation to h is hard to predict because it is determined by the total sum of the power consumption per unit SOA, the number of SOAs in the network and the laser source power. Simulation results show that the power consumption is the largest when $h = 0$; therefore, placing SOAs on all links is inefficient in terms of power consumption. Contrarily, the minimum power consumption is achieved when $h = 2$ and the overall power consumption of the SOA-enabled HONoC is reduced by 32.17–45.49%, compared to the conventional HONoC.

Figure 18 shows the comparison of power consumption between the SOA-enabled HONoC and the conventional HONoC according to the mesh size and the hop count between SOAs in a random uniform pattern. The increase in the power consumption in the proposed HONoC is relatively low compared to that in the conventional HONoC because the insertion loss is independent of the longest path. As a result, the power saving realized by the proposed approach becomes prominent as the mesh size grows.

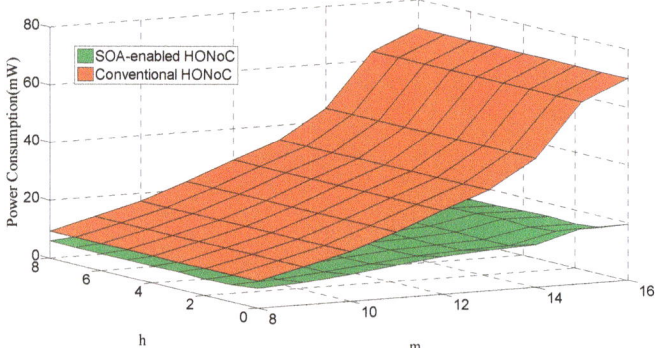

Figure 18. Power consumption comparison between a conventional HONoC (without SOA) and SOA-enabled HONoC in terms of mesh size and maximum hop count without SOAs.

6. Conclusions

We proposed the SOA-enabled HONoC architecture to reduce the total power consumption and mitigate SNR degradation, considering network size scalability without increasing laser source output power. An SOA placement algorithm was developed for efficient laser source power reduction and SNR enhancement. We suggested the worst-crosstalk noise model and presented the attuning method for optimal SOA gain with respect to power consumption and SNR, respectively. We constructed a C-based simulator to evaluate the SOA-enabled HONoC with an associated SOA placement algorithm. Simulation results showed that the worst-case SNRs for the 1st, 2nd and 3rd longest paths are improved by an average of 4.26 dB and power consumption is reduced by 32.17% to 45.49% under various traffic patterns in the 8×8 network. The performance gaps regarding various SOA spacings were not significant; on the other hand, the improvement achieved in both SNR and power consumption was increased in proportion to the mesh size. Therefore, the proposed SOA-enabled HONoC can be a scalable solution to cope with the performance degradation problem of large-scale HONoCs.

Author Contributions: Formal analysis, J.Y.J.; Investigation, C.-L.L.; Methodology, M.S.K.; Validation, J.Y.J.; Writing—original draft, J.Y.J.; Writing—review & editing, T.H.H.

Funding: This research was funded by the Basic Science Research Program through the National Research Foundation of Korea under Grant NRF-2018R1D1A1B07043585 and by the MOTIE (Ministry of Trade, Industry & Energy) (10080594) and KSRC (Korea Semiconductor Research Consortium) support program for the development of the future semiconductor device.

Conflicts of Interest: The authors declare no conflict of interest.

References

1. Vantrease, D.; Schreiber, R.; Monchiero, M.; McLaren, M.; Jouppi, N.P.; Fiorentino, M.; Davis, A.; Binkert, N.; Beausoleil, R.G.; Ahn, J.H. Corona: System implications of emerging nanophotonic technology. In Proceedings of the IEEE/ACM International Symposium on Computer Architecture (ISCA), Beijing, China, 21–25 June 2008.
2. International Technology Roadmap for Semiconductors. Available online: http://www.itrs2.net (accessed on 27 September 2018).
3. Shacham, A.; Bergman, K.; Carloni, L.P. Photonics Networks-on-Chip for Future Generations of Chip Multiprocessors. *IEEE Trans. Comput.* **2008**, *57*, 1246–1260. [CrossRef]

4. Mo, K.H.; Ye, Y.; Wu, X.; Zhang, W.; Liu, W.; Xu, J. A Hierarchical Hybrid Optical-Electronic Network-on-Chip. In Proceedings of the 2010 IEEE Computer Society Annual Symposium on VLSI, Lixouri, Kefalonia, Greece, 5–7 July 2010.
5. Gu, H.; Xu, J.; Wang, Z. A novel optical mesh network-on-chip for gigascale systems-on-chip. In Proceedings of the 2008 IEEE Asia Pacific Conference on Circuits and Systems, Macao, China, 30 November–3 December 2008.
6. Lee, J.; Kim, Y.; Li, C.; Han, T. A shortest path adaptive routing technique for minimizing path collision in hybrid optical network-on-chip. *J. Syst. Arch.* **2013**, *59*, 1334–1347. [CrossRef]
7. Lan, F.; Wu, R.; Zhang, C.; Pan, Y.; Cheng, K.-T.T. DLPS: Dynamic laser power scaling for optical Network-on-Chip. In Proceedings of the 2017 22nd Asia and South Pacific Design Automation Conference (ASP-DAC), Chiba, Japan, 16–19 January 2017.
8. Xie, Y.; Nikdast, M.; Xu, J.; Zhang, W.; Li, Q.; Wu, X.; Ye, Y.; Wang, X.; Liu, W. Crosstalk noise and bit error rate analysis for optical network-on-chip. In Proceedings of the ACM/IEEE Design Automation Conference, Anaheim, CA, USA, 13–18 June 2010.
9. Xiao, S.; Khan, M.H.; Shen, H.; Qi, M. Multiple-channel silicon micro-resonator bad filters for WDM applications. *Opt. Express* **2007**, *15*, 7489–7498. [CrossRef] [PubMed]
10. Eid, N.; Boeck, R.; Jayatilleka, H.; Chrostowski, L.; Shi, W.; Jaeger, N.A.F. A silicon-on-insulator microring resonator filter with bent contradirectional couplers. In Proceedings of the 2016 IEEE Photonics Conference (IPC), Waikoloa, HI, USA, 2–6 October 2016.
11. Fusella, E.; Cilardo, A. Lighting Up On-Chip Communications with Photonics: Design Tradeoffs for Optical NoC Architectures. *IEEE Circuits Syst. Mag.* **2016**, *16*, 4–14. [CrossRef]
12. Poon, A.W.; Xu, F.; Luo, X. Cascaded active silicon microresonator array cross-connect circuits for WDM networks-on-chip. In Proceedings of the SPIE—The International Society for Optical Engineering, San Jose, CA, USA, 13 February 2008.
13. Ding, W.; Tang, D.; Liu, Y.; Chen, L.; Sun, X. Compact and low crosstalk waveguide crossing using impedance matched metamaterial. *Appl. Phys. Lett.* **2010**, *96*, 111114. [CrossRef]
14. Biberman, A.; Preston, K.; Hendry, G.; Sherwood-Droz, N.; Chan, J.; Levy, J.S.; Lipson, M.; Bergman, K. Photonic network-on-chip architectures using multilayer deposited silicon materials for high-performance chip multiprocessors. *ACM J. Emerg. Technol. Comput. Syst.* **2011**, *7*, 7. [CrossRef]
15. Dong, P.; Qian, W.; Liao, S.; Liang, H.; Kung, C.-C.; Feng, N.-N.; Shafiiha, R.; Fong, J.; Feng, D.; Krishnamoorthy, A.V.; et al. Low Loss Silicon Waveguides for Application of Optical Interconnects. In Proceedings of the IEEE Photonics Society Summer Topicals 2010, Playa del Carmen, Mexico, 19–21 July 2010.
16. Lee, B.G.; Biberman, A.; Dong, P.; Lipson, M.; Bergman, K. All-Optical Comb Switch for Multiwavelength Message Routing in Silicon Photonic Networks. *IEEE Photonics Technol. Lett.* **2008**, *20*, 767–769. [CrossRef]
17. Sanchis, P.; Galan, J.V.; Brimont, A.; Griol, A.; Marti, J.; Piqueras, M.A.; Perdigues, J.M. Low-crosstalk in silicon-on-insulator waveguide crossings with optimized-angle. In Proceedings of the 2007 4th IEEE International Conference on Group IV Photonics, Tokyo, Japan, 19–21 September 2007.
18. Chen, H.; Poon, A.W. Low-Loss Multimode-Interference-Based Crossings for Silicon Wire Waveguides. *IEEE Photonics Technol. Lett.* **2006**, *18*, 2260–2262. [CrossRef]
19. Xie, Y.; Nikdast, M.; Xu, J.; Wu, X.; Zhang, W.; Ye, Y.; Wang, X.; Wang, Z.; Liu, W. Formal Worst-Case Analysis of Crosstalk Noise in Mesh-Based Optical Networks-on-Chip. *IEEE Trans. Very Large Scale Integr. Syst.* **2013**, *21*, 1823–1836. [CrossRef]
20. Nikdast, M.; Xu, J.; Wu, X.; Zhang, W.; Ye, Y.; Wang, X.; Wang, Z.; Wang, Z. Systematic Analysis of Crosstalk Noise in Folded-Torus-Based Optical Networks-on-Chip. *IEEE Trans. Comput. Aided Des. Integr. Circuits Syst.* **2014**, *33*, 437–450. [CrossRef]
21. Duong, L.H.K.; Nikdast, M.; Xu, J.; Wang, Z.; Thonnart, Y.; Beux, S.L.; Yang, P.; Wu, X.; Wang, Z. Coherent crosstalk noise analyses in ring-based optical interconnects. In Proceedings of the 2015 Design, Automation and Test in Europe Conference and Exhibition (DATE), Grenoble, France, 9–13 March 2015.
22. Nikdast, M.; Xu, J.; Duong, L.H.K.; Wu, X.; Wang, Z.; Wang, X.; Wang, Z. Fat-Tree-Based Optical Interconnection Networks Under Crosstalk Noise Constraint. *IEEE Trans. Very Large Scale Integr. Syst.* **2014**, *23*, 156–169. [CrossRef]

23. Rostami, A.; Baghban, H.; Maram, R. *Nanostructure Semiconductor Optical Amplifiers*, 1st ed.; Springer: Berlin, Germany, 2011; Available online: https://books.google.com.hk/books?hl=en&lr=&id=xN0PXG_38SIC&oi=fnd&pg=PR3&dq=Nanostructure+Semiconductor+Optical+Amplifiers&ots=A-G24LLZxf&sig=QA-Sb2u1vHjRxW_K80vxRLDUB4E&redir_esc=y#v=onepage&q=Nanostructure%20Semiconductor%20Optical%20Amplifiers&f=false (accessed on 16 September 2018).
24. Thakkar, I.G.; Chittamuru, S.V.R.; Pasricha, S. Run-time laser power management in photonic NoCs with on-chip semiconductor optical amplifiers. In Proceedings of the 2016 Tenth IEEE/ACM International Symposium on Networks-on-Chip, Nara, Japan, 31 August–2 September 2016.
25. Ye, Y.; Wu, X.; Xu, J.; Zhang, W.; Nikdast, M.; Wang, X. Holistic comparison of optical routers for chip multiprocessors. In Proceedings of the IEEE 2012 International Conference on Anti-Counterfeiting, Security and Identification (ASID), Taipei, Taiwan, 24–26 August 2012.
26. Jiang, N.; Becker, D.U.; Michelogiannakis, G.; Balfour, J.; Towles, B.; Shaw, D.E.; Kim, J.; Dally, W.J. A detailed and flexible cycle-accurate Network-on-Chip simulator. In Proceedings of the 2013 IEEE International Symposium on Performance Analysis of Systems and Software (ISPASS), Austin, TX, USA, 21–23 April 2013.
27. Bogaerts, W.; Dumon, P.; Thourhout, D.V.; Baets, R. Low-loss, low-cross-talk crossings for silicon-on-insulator nanophotonic waveguides. *Opt. Lett.* **2007**, *32*, 2801–2803. [CrossRef] [PubMed]
28. Chan, J.; Hendry, G.; Bergman, K.; Carloni, L.P. Physical-Layer Modeling and System-Level Design of Chip-Scale Photonic Interconnection Networks. *IEEE Trans. Comput. Aided Des. Integr. Circuits Syst.* **2011**, *30*, 1507–1520. [CrossRef]
29. Maulini, R.; Lyakh, A.; Tsekoun, A.; Patel, C.K.N. λ∼7.1 µm quantum cascade lasers with 19% wall-plug efficiency at room temperature. *Opt. Express* **2011**, *19*, 17203–17211. [CrossRef] [PubMed]
30. Joshi, A.; Batten, C.; Kwon, Y.-J.; Beamer, S.; Shamim, I.; Asanovic, K.; Stojanovic, V. Silicon-photonic clos networks for global on-chip communication. In Proceedings of the 2009 3rd ACM/IEEE International Symposium on Networks-on-Chip, San Diego, CA, USA, 10–13 May 2009.

© 2019 by the authors. Licensee MDPI, Basel, Switzerland. This article is an open access article distributed under the terms and conditions of the Creative Commons Attribution (CC BY) license (http://creativecommons.org/licenses/by/4.0/).

Article

Two-Dimensional Constellation Shaping in Fiber-Optic Communications

Zhen Qu [1],*, Ivan B. Djordjevic [2] and Jon Anderson [1]

[1] Juniper Networks, 1133 Innovation Way, Sunnyvale, CA 94089, USA; jonanderson@juniper.net
[2] Department of Electrical and Computer Engineering, University of Arizona, 1230 E. Speedway Blvd., Tucson, AZ 85721, USA; ivan@email.arizona.edu
* Correspondence: zqu@juniper.net; Tel.: +1-520-442-7197

Received: 5 April 2019; Accepted: 2 May 2019; Published: 8 May 2019

Abstract: Constellation shaping has been widely used in optical communication systems. We review recent advances in two-dimensional constellation shaping technologies for fiber-optic communications. The system architectures that are discussed include probabilistic shaping, geometric shaping, and hybrid probabilistic-geometric shaping solutions. The performances of the three shaping schemes are also evaluated for Gaussian-noise-limited channels.

Keywords: constellation shaping; probabilistic shaping; geometric shaping; fiber-optic communications; quadrature amplitude modulation; mutual information; generalized mutual information

1. Introduction

The advances in photonics integrated circuit [1,2], software-defined networking [3,4], application-oriented fibers [5,6], and optical amplifiers [7,8], have been greatly pushing forward development and commercialization of optical communications. In the modern optical transport systems, especially in terrestrial and transoceanic fiber-optic communications, advanced modulation formats have been overwhelmingly implemented in optical transceivers [9–14], thanks to the ever-cheaper optical frontend and powerful digital signal processing (DSP) chips with smaller size and lower power consumption. Traditional quadrature amplitude modulation (QAM) formats have been applied extensively to realize high-capacity and long-reach optical communications, and we have witnessed numerous two-dimensional QAM (2D-QAM)-based hero experiments in recent years to explore the highest possible system capacity and longest transmission distance [15–17].

Due to the loss profile of the standard single mode fiber (SSMF) and gain profile of commercial Erbium-doped fiber amplifiers (EDFAs), C-band window (1530–1565 nm) is mostly used for data loading [18]. However, the capacity bottleneck becomes more visible for the traditional QAM-based C-band optical transmissions [19]. In order to meet the increasing bandwidth requirements, in particular the upcoming 5G infrastructure, more advanced solutions are expected to be introduced to optical infrastructures. We can roughly divide the promising solutions into two categories: coded modulation [20–27] and extended multiplexing [17,28–31]. The idea of coded modulation is increasing information bits per channel use, including higher-order 2D modulation formats, like 1024QAM [22], multidimensional QAM formats, like 4D optimized constellations [23], and constellation shaping techniques [24–26], like probabilistic shaping (PS)-64QAM [25]. Alternatively, extended multiplexing is a more straightforward solution, which mainly contains C+L band wavelength multiplexing [17,28] and space-division multiplexing [29–31]. All the solutions come with the trade-off choice between optical complexity and electrical complexity. In order to implement C+L band wavelength multiplexing, C-band EDFAs and L-band EDFAs should be used together to simultaneously amplify the channel loss per span [17]. Raman amplifier may be another option [32], but the telecom industry is not in favor of it because of its high cost. Space-division multiplexing based on few-mode fiber or multi-core fiber is an efficient way to directly boost the aggregate capacity. However, it seems to have been put at the bottom of the to-do list, because (i) it will be quite challenging to replace and rebuild the current fiber links, (ii) space-division-multiplexing-based optical amplifiers are too expensive for commercial applications, and (iii) complex multi-input multi-output (MIMO) channel equalization may be required to compensate channel crosstalk [33]. The telecom industry always chooses the most cost-efficient way to upgrade their communication systems. Therefore, extended multiplexing solution is barely found in the roadmap of optical communications, especially in the field of long-reach optical communications. On the contrary, coded modulation seems to be a more attracting solution. Higher modulation formats are more common now, but they have more stringent requirements on optical signal-to-noise ratio (OSNR), digital-to-analog converter/analog-to-digital converter (DAC/ADC), and DSP recovery. Multidimensional QAM solution can further enlarge the minimum Euclidean distance between constellation points, but it requires powerful DSP technology to recover signals. What is more, such performance optimization can come at a price of less information bits per channel use [34].

In traditional QAM formats, the constellation points are located on a uniform grid. Such uniform QAM formats are easy for generation and recovery but suffer a 1.53-dB asymptotic loss towards the Shannon limit. In order to close the gap, constellation shaping was introduced to optical communications. Constellation shaping, including PS [25,35–37], geometric shaping (GS) [26,38–41], and hybrid probabilistic-geometric shaping (HPGS) [42–45], is used to mimic a Gaussian distribution with limited constellation points. Although target at approaching Gaussian distribution, PS and GS have completely different generation and detection implementations. GS-QAM is obtained by optimizing some metrics, like mutual information (MI) [40] and generalized MI (GMI) [41], which will relocate the constellation point in geometric space. PS-QAM is applied on a uniform grid, but the constellation points will be transmitted with different probabilities. HPGS can optimize the performance in both geometric space and probabilistic space which should, in principle, provide the optimal performance.

In this paper, we review recent advances on 2D constellation shaping in fiber-optic communications. Section 2 gives an overview of PS, GS, and HPGS. Section 3 discusses the performance of PS, GS, and HPGS in Gaussian-noise-limited channels. Finally, the concluding remarks are given in Section 4.

2. Typical Constellation Shaping Schemes

2.1. Probabilistic Shaping

In a Gaussian-noise-limited channel, PS-QAM yielding a Maxwell–Boltzmann (M–B) distribution is recognized as the optimal format to maximize the channel capacity [46]. In general, the low-amplitude constellation points are sent with a higher probability than the high-amplitude ones. Besides, the constellation points under the same amplitude layer are sent equally likely. Therefore, the average symbol power will be decreased, but at a cost of lower source entropy.

The first PS scheme was proposed by Gallager, which is based on many-to-one mapping [47]. Complex forward error correction (FEC) coding is required to be implemented to recover the original bits from the systematic errors after many-to-one demapping. The recently proposed arithmetic coding-based constant composition distribution matcher (CCDM) is an invertible fix-to-fix length distribution matcher, enabling maximum information rate asymptotically in the block-length [48]. Later, there are some methods proposed to further reduce the complexity of CCDM, like multiset-partition distribution matching [49] and streaming distribution matching [50]. The first PS-based coded modulation was probabilistic amplitude shaping (PAS) [51], which could seamlessly combine the binary FEC coding and CCDM in a square M-QAM format.

The proposed PAS enables capacity approaching fiber-optic communications, but also brings some issues. Firstly, CCDM and the modified CCDM architectures are hard to be implemented in commercial optical transceivers. Secondly, more bit-to-symbol (B2S) mapping and symbol-to-bit (S2B) mapping modules are required to be implemented at the transceivers, leading to extra complexity.

Thirdly, there will be an entropy loss by shaping on a given M-ary signal constellation format. Fourthly, intrablock error propagation will loom over the dematching procedure once the FEC coding cannot totally correct all bit errors. Fifthly, the applicable FEC code rate is limited, which will be lower-bounded by $[\log_2(M) - 2] / \log_2(M)$. Sixthly, DSP circuit will be under a higher pressure to recover Gaussian-like constellation diagrams. Thereafter, more pilot-tones are essential to be used for signal recovery. Last but not least, although could be used for reach extension, PAS-MQAM suffers more from the modulation-dependent noise in long-haul fiber-optic communications [52].

In a square MQAM format, the coordinate of each constellation can be represented by the Cartesian product of two pulse-amplitude modulation (PAM) coordinates, namely,

$$X = \left\{\pm 1, \pm 3, \ldots, \pm\left(\sqrt{M} - 1\right)\right\} \tag{1}$$

The well-known M–B distribution is defined by

$$P_{X_v}(x) = e^{-v|x|^2} / \sum_{x' \in X} e^{-v|x'|^2} \tag{2}$$

where v is a non-negative scaling factor. If v is 0, the PAM format follows a uniform distribution.

The capacity of PAS-MQAM format can be defined as [51]

$$C = H(p) - m(1 - R) \tag{3}$$

where $H(p)$ is the entropy of the PAS-MQAM format, R is the FEC code rate, and $m = \log_2(M)$.

Figure 1 shows the conceptual diagram of PAS-MQAM generation, where 16QAM is used as an illustrative example. In such scheme, the 1D amplitude symbols labeled by 1-3-1-1-1-3 ..., will be probabilistically shaped according to the M–B distribution (the probabilities of Symbol-1 and Symbol-3 are indicated by the blue and black colors, respectively), and the sign bits labeled by 1-0-0-1-0-1 ..., will be used to carry the uniformly distributed parity-check bits. The S2B mapping is used to map the Symbol-1 and Symbol-3 to amplitude bits, i.e., Bit-1 and Bit-0, respectively. In the FEC encoder, the party-check bits will be appended to the sign bits, and combine with the amplitude bits. Such encoded bits are remapped to PAM-4 symbols (00→−3, 01→−1, 11→+1, 10→+3). Therefore, after FEC coding, the M–B distribution will not be changed, and Gray-mapping rule is still applicable.

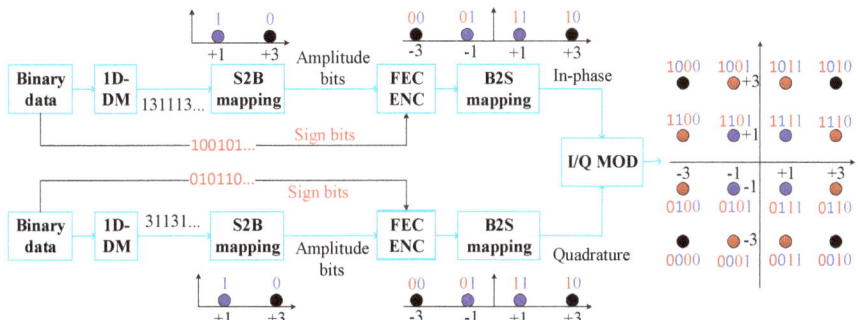

Figure 1. The conceptual diagram of probabilistic amplitude shaping (PAS)-M-ary quadrature amplitude modulation (MQAM) generation.

2.2. Geometric Shaping

GS-QAM is generated by optimizing certain criterion under a given signal-to-noise ratio (SNR). Such criteria can be maximizing MI [40], maximizing GMI [41], maximizing constellation figure of merit [53], or minimizing mean-square error of Gaussian distribution [54], etc. Typically, the lower-amplitude constellation points will be more concentrated around the origin. Generalized cross constellations and Voronoi constellations were proposed decades ago [38,39]. GMI-optimized constellations have also been proposed recently to enhance the capacity in a binary FEC coding featured fiber-optic communication systems.

GS could potentially simplify the process to generate Gaussian-like constellations, but it also brings several issues. Firstly, due to the unavailability of Gray-mapping in most cases, GMI performance hardly approach MI performance. Although nonbinary FEC coding is a solution to close such gap, the high complexity hinders its commercial development [55]. Secondly, the common asymmetry of the constellations may result in more complex PAM constellations in both in-phase and quadrature branches. Therefore, the ADC/DAC is required to be implemented with higher resolutions. Thirdly, the DSP circuits to recover GS-MQAM formats are not compatible with the ones to recover regular-MQAM. As a result, new DSP algorithms are suggested to be developed to efficiently recover the GS-MQAM signals. Fourthly, it is hard to reach standard agreements between optical transceiver manufacturers due to the variety of GS-QAM formats and the matched DSP algorithms.

Figure 2 shows the conceptual diagram of a GS-QAM-based communication system. If binary FEC coding is used, the suboptimal B2S mapping table has to be found to minimize the gap between GMI and MI, in order to minimize the Non-Gray mapping penalty. Brutal force algorithm may be used to find such mapping rule at a cost of high computational complexity. Alternatively, binary data can be mapped to the GS-QAM symbols through any B2S look-up table, followed by a reasonable complexity nonbinary FEC encoder to sustain reliable communications [56].

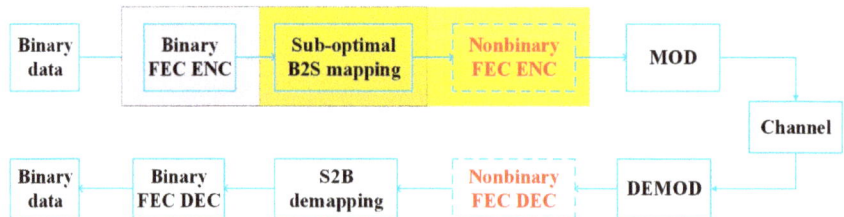

Figure 2. The conceptual diagram of a geometric shaping (GS)-QAM-based communication system.

Figure 3 shows 16/32/64QAM formats generated by maximizing constellation figure of merit [53], minimizing mean-square error of Gaussian distribution [57], and maximizing GMI [41], respectively. The GS-16QAM constellation shown in Figure 3a is obtained by maximizing the minimum Euclidean distance under the same average power of the 16-ary constellation.

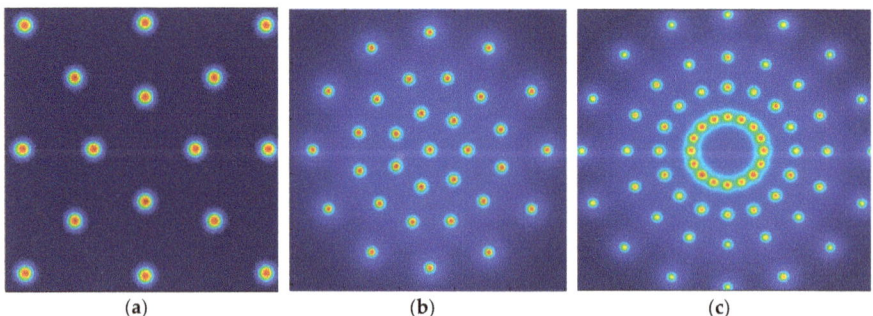

Figure 3. GS-QAM formats based on (**a**) maximizing constellation figure of merit, (**b**) minimizing mean-square error of Gaussian distribution, and (**c**) maximizing generalized mutual information (GMI).

The GS-32QAM constellation shown in Figure 3b can be obtained by the following steps.

1. Choosing 2D Gaussian distribution as the optimal source distribution, and select the uniformly distributed regular-32QAM as the initial constellation.
2. Generating a symbol sequence following the Gaussian distribution.
3. Distributing the symbols into 32 clusters, where the decision is made as per the minimum Euclidean distance from the 32QAM constellation points obtained in previous iteration.
4. Finding the average central positions from the symbols labeled by each cluster. Such 32 points located on the central positions are used as the new MQAM constellation points.
5. Iterating over Steps 2 and 4 until convergence.

The GS-64QAM constellation shown in Figure 3b is designed with the constraint of Gray mapping, which can maximize the GMI. Due to the Gray mapping constraint, such scheme cannot fully explore the 2D space, but it can also reduce capacity gap towards the Shannon limit.

It is an open question about the optimal GS solution, since it depends on lots of implementation scenarios. When the amount of the constellation points is more than 64, GS may suffer more implementation penalties than PS.

2.3. Hybrid Probabilistic-Geometric Shaping

When more flexible constellation formats are required, especially multi-dimensional QAM formats, HPGS may be an enabling technique, since each constellation point during the optimization process will not be limited to equal probability or uniform-grid locations. Figure 4 shows HPGS-5QAM and HPGS-9QAM formats based on Huffman coding [58,59]. The probability of each symbol is predetermined by the corresponding Huffman tree, and the coordinates of the symbols can be obtained by optimizing the MI.

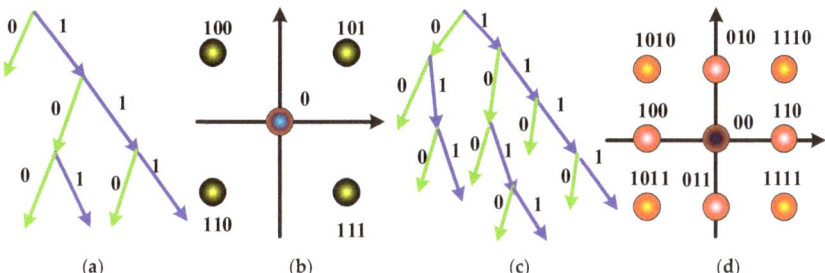

Figure 4. Huffman coding (**a**) and the constellation format (**b**) for 5-QAM; Huffman coding (**c**) and the constellation format (**d**) for 9-QAM.

As an illustrative example, the generation process of HPGS-9QAM constellation can be found below:

1. Parsing the binary source into nine blocks labeled by unique bit sets {00, 010, 110, 011, 100, 1110, 1111, 1010, 1011}. If the binary sequence is sufficiently long, the resulting blocks should be generated with the probabilities of {$P(00) = 1/4$, $P(010) = 1/8$, $P(011) = 1/8$, $P(100) = 1/8$, $P(1110) = 1/16$, $P(1111) = 1/16$, $P(1010) = 1/16$, $P(1011) = 1/16$}. Thereafter, the entropy is 3, i.e., there is no entropy loss.
2. Mapping the 9-block sequence to any 9-QAM symbols with the constraints: (i) The 9-ary constellation points with the same probability are uniformly located in the same power layer, (ii) The constellation points with higher probabilities are located at higher power layers. In other words, such 9-ary constellation should be featured with three power layers and 1, 4, and 4 points are equally spaced in each power layer, respectively.

3. Maximizing the MI by iterating over all amplitude ratios and phase differences of such 9-ary constellation.

Huffman coding can be treated as a variable-length and prefix-free PS scheme, which can also be uniquely decodable, but it cannot provide flex rate.

There are four major problems for this HPGS scheme. Firstly, it is quite challenging to implement Huffman coding when the amount of symbols increase. Secondly, as any other variable-length coding technology, Huffman coding will also suffer from the overflow or underflow problems. Thirdly, the higher-complexity nonbinary FEC coding is required. Fourthly, error propagation may occur if unexpected symbol errors remained after FEC decoding.

A more efficient and practical way to generate HPGS-QAM is based on universal probabilistic shaping scheme and GMI-optimized GS scheme [44,60]. Figure 5 shows the constellation formats for HPGS-32QAM, as well as the regular/PS-32QAM.

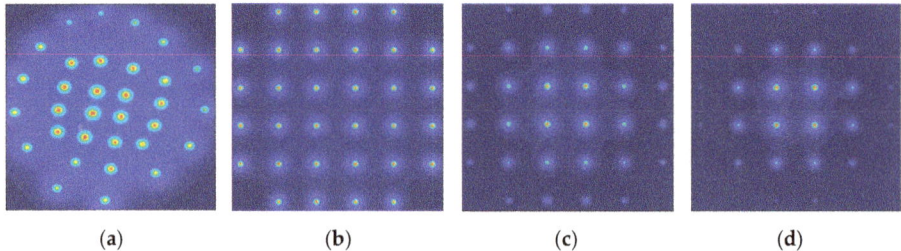

Figure 5. Constellation formats for (**a**) hybrid probabilistic-geometric shaping (HPGS)-32QAM, (**b**) regular-32QAM, (**c**) shallowly shaped 32QAM, and (**d**) deeply shaped 32QAM.

The constellation diagram of the HPGS-32QAM shown in Figure 5a is generated by the generalized pairwise optimization algorithm. The objective function is maximizing GMI under the constraints of zeros mean amplitude and normalized average power. The two constraints can be expressed as

$$p_i x_i = -A - p_j x_j \qquad (4)$$

$$\left|p_j x_j + b\right|^2 + \frac{p_j}{p_i^2}|x_j|^2 = 1 - B \qquad (5)$$

where (x_i, x_j) is one pair of the M-ary constellation ($M = 32$ in this case), p_i is the probability of x_i, $A = \sum_{k=1, k \neq i, k \neq j}^{M} p_k x_k$, and $B = \sum_{k=1, k \neq i, k \neq j}^{M} p_k |x_k|^2$. The objective of maximizing GMI will be searching the (x_i, x_j) pair over a hypersphere with the center and radius determined by the other $M - 2$ constellation points. Such generalized pairwise optimization algorithm can converge to the final steady state after iterating all $M(M-1)/2$ pairs. As a rule of thumb, we suggest start the iteration with the regular-32QAM, and the optimal HPGS-32QAM shown in Figure 5a can be obtained within 1000 iterations.

Such HPGS constellation could provide the trade-off between the number of nearest symbols and their Hamming distance. Any constellation format shown in Figure 5 does not belong to the square QAM category. As a result, the well-known PAS scheme cannot be used for the PS purpose.

Figure 6 shows the modified probabilistic fold shaping and universal probabilistic shaping schemes [60]. Probabilistic fold shaping can be used for shaping any F-fold rotationally symmetric constellation, like regular-32QAM, as shown in Figure 6a. In such kind of constellations, the bits determining the fold index yield uniform distribution, which can be used to carry the parity-check bits. There is a major difference between PAS and probabilistic fold shaping. PAS is a 1D shaping scheme, which uses the single bit determining the positive or negative amplitude to carry the parity-check bit and shapes the 1D-PAM with a 1D M–B distribution. On the contrary, probabilistic fold shaping is a 2D shaping scheme, which uses the $\log_2(F)$ bits determining the fold indexes to carry the parity-check bits, and shapes the 2D constellation points in one fold with a 2D M–B distribution. As an illustrative example, the 32QAM shown in Figure 6a is featured with four-fold rotational symmetry. The 8-ary constellation points in one fold are firstly shaped with a 2D M–B distribution, where different colors indicate different probabilities. The bit sets {11,01,00,10} determining the fold indexes will carry the parity-check bit after FEC encoding and rotate the 8-ary constellation by 0°, 90°, 180°, 270°, respectively. Therefore, 2D M–B distribution can be applied to the constellation points in one fold, and the selection of the fold-index can be performed by the parity-check bits. The target distribution will not be changed after the binary FEC coding.

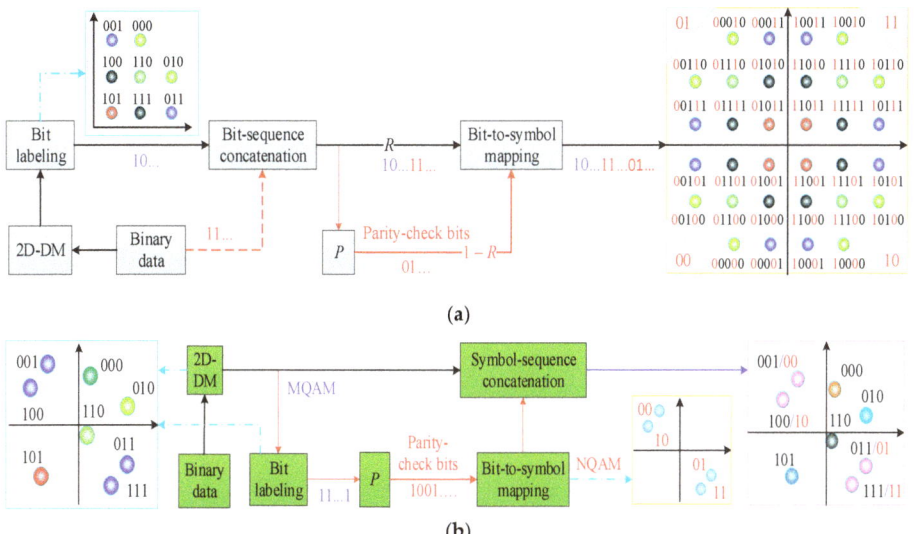

Figure 6. Modified probabilistic shaping (PS) schemes based on (**a**) probabilistic fold shaping, (**b**) universal PS.

The universal probabilistic shaping scheme shown in Figure 6b can be applied to shape any QAM format. The MQAM symbols generated from the CCDM may not yield a M–B distribution. The binary bits generated after the bit labeling block are used to carry the uniformly distributed parity-check bits. During the process of B2S mapping, the parity-check bits will be uniformly mapped to partial MQAM symbols, i.e., N-ary QAM (NQAM) symbols, where N is the largest power of 2 to contain the MQAM constellation points with the desirable probabilities of $>(1-R)/M$. Assuming that the desirable probability distribution of the MQAM symbols is $P_M(x)$, the probability distribution of the MQAM symbols after the CCDM is $P_D(x)$, and probability distribution of the NQAM symbols is given by $P_N(y) = 1/N$. The final relationship can be written as $P_M(x) = \left[RP_D(x)/\log_2(M) + (1-R)P_N(y)/\log_2(N)\right]/\left[R/\log_2(M) + (1-R)/\log_2(N)\right]$. In such a way, any desirable distribution of the HPGS-MQAM symbols can be obtained.

3. Performance Comparison in Gaussian-Noise-Limited Channels

As a rule of thumb, HPGS-MQAM cannot show clear performance improvement over PS-MQAM and GS-MQAM when M < 32. In addition, if M ≥ 64, PS-MQAM could closely approach the Shannon limit, thus it is not necessary to apply HPGS to higher order QAM formats. In this paper, we performed Monte Carlo simulations in MATLAB. The block-length of the CCDM was chosen more than 5000, so the normalized divergence of the encoder output and the desired distribution is negligible. The *awgn* function provided by MATLAB was used to add white Gaussian noise to the 2D-MQAM signals. All the results were averaged over 1000 trials.

In a numerical simulation, as shown in Figure 7, we compared the MI performances of the minimizing mean-square error of Gaussian-distribution-based GS-8/16/32QAM, CCDM-based PS-8/16/32QAM, and regular-8/16/32QAM. Figure 7a1,a2,b1,b2,c1,c2 show the constellation diagrams of PS/GS-8/16/32QAM. Here we chose MI as the metric for performance comparison, because MI can be measured without the consideration of the FEC coding performance. Given that most of the current nonbinary FEC coding and binary FEC coding schemes may vary in performance and implementation penalty, MI instead shows the upper limit of the capacity obtained with the "ideal" FEC coding scheme [61]. In order to easily describe the PS-MQAM with an entropy of *A* b/s, we adopt the notation of PS-MQAM*A*. For example, we use PS-8QAM2.3 to denote the PS-8QAM with an entropy of 2.3 b/s. As we can see from Figure 7a,b, the best MI performances can be obtained by GS-8/16QAM. GS-8/16QAM can always outperform regular-8/16QAM. PS-8QAM can have comparable performance over GS-8QAM when the SNR is less than 6.2 dB; the performances of PS-16QAM and GS-16QAM are quite similar when the SNR is less than 11.7 dB. In the region of high SNR region, PS-8/16QAM cannot bring better MI performance. In Figure 7c, we find that the best performance can be achieved by PS/GS-32QAM separately. In addition, GS-32QAM is always better than regular-32QAM in terms of MI performance. When the SNR is more than 15.7 dB, GS-32QAM has the best performance, while PS-32QAM formats can maximize the MI performance when the SNR is less than 15.7 dB.

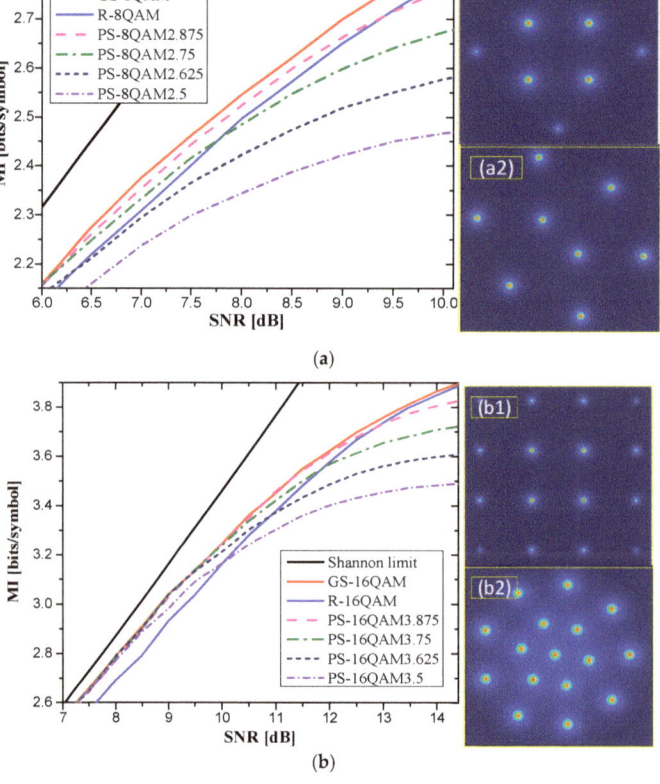

Figure 7. *Cont.*

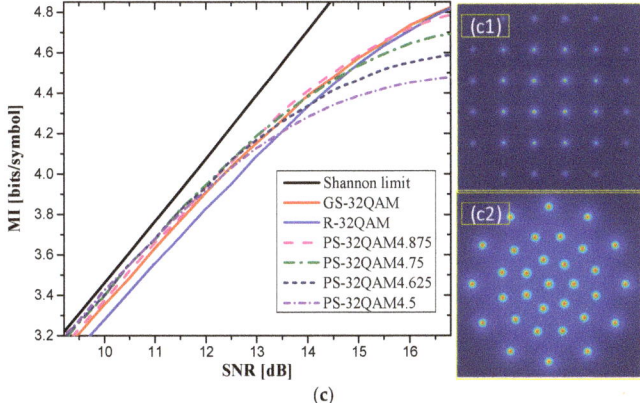

(c)

Figure 7. Mutual information (MI) versus signal-to-noise ratio (SNR) performance in Gaussian-noise-limited channels for (**a**) PS/GS/R-8QAM, (**b**) PS/GS/R-16QAM, (**c**) PS/GS/R-32QAM. Insets (a1,a2) The constellation diagrams of PS-8QAM and GS-8QAM. Insets (b1,b2) The constellation diagrams of PS-16QAM and GS-16QAM. Insets (c1,c2) The constellation diagrams of PS-32QAM and GS-32QAM.

Although PS-256QAM and PS-1024QAM have been investigated over fiber links [62], we still believe that PS-64QAM is expected to be the most promising solution for high-order QAM-based fiber-optic communications, at least for the upcoming 400 G/800 G Ethernet. It is mainly because that PS-64QAM has near-capacity-approaching performance, and its implementation penalty may also be well reduced to an acceptable level in the near future. While, unfortunately, regular-64QAM and PS-64QAM have been suffering a large implementation penalty so far. Here we compare the performances of PAS-64QAM and GMI-optimized HPGS-32QAM (referred to as opti-32QAM in this paper).

In another numerical simulation, Figure 8 shows the post-FEC bit error rate (BER) versus SNR performances of PAS-64QAM, opti-32QAM, PS-32QAM, and regular-32QAM. DVB-S2 irregular low-density parity-check (LDPC) codes were applied for FEC coding. The performance comparisons were executed under the same capacity levels, i.e., C = 3.33 b/s and C = 4 b/s. Table 1 lists the parameters used in the post-FEC BER analysis under the capacity levels of 3.33 b/s and 4 b/s. In Figure 8a, the performance of PS-32QAM is slightly better than opti-32QAM, but worse than PAS-64QAM by 0.2 dB. Opti-32QAM shows a 0.8-dB performance improvement over regular-32QAM in case of C = 3.33 b/s. In Figure 8b, the performance of opti-32QAM is better than PS-32QAM by 0.2 dB, better than regular-32QAM by 0.8 dB when the capacity is 4 b/s. However, PAS-64QAM is also shown to outperform opti-32QAM by 0.4 dB. It is reasonable to find that PAS-64QAM always has the best post-FEC performance over shaped 32QAM. While in a realistic communication system, we believe that the performance of HPGS-32QAM (opti-32QAM) should be similar to that of PAS-64QAM, due to the higher implementation penalties that PAS-64QAM may suffer.

Table 1. The parameters used in the post-forward error correction (FEC) bit error rate (BER) analysis under the same capacity.

C [b/s]	H/R	R-32QAM	Opti-32QAM	PS-32QAM	PAS-64QAM
3.33	$H(p)$	5	5	4.33	4.53
	R	2/3	2/3	4/5	4/5
4	$H(p)$	5	5	4.55	5.2
	R	4/5	4/5	8/9	4/5

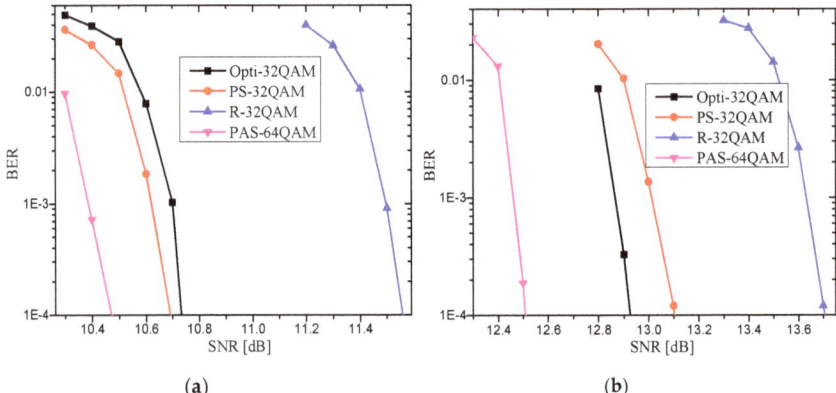

Figure 8. BER versus SNR performances in the scenarios of (**a**) C = 3.33 b/s, (**b**) C = 4 b/s.

4. Concluding Remarks

Constellation shaping will play an increasingly important role in fiber-optic communications in the wake of the booming 5G era. In this paper, we focused on the performance of 2D constellation shaping in Gaussian-noise-limited channels. We have discussed three key constellation shaping schemes, i.e., PS, GS, and HPGS, and analyzed their pros and cons in terms of performance and implementation complexity. We also introduced two modified CCDM-based shaping schemes, i.e., probabilistic fold shaping and universal PS, which could enable applying PS on any 2D modulation format. We found that GS-8QAM and GS-16QAM could outperform PS/regular-8QAM and PS-regular-16QAM, respectively, in terms of MI performance. In addition, the best MI performance of 32-ary QAM format could be reached by PS-32QAM and GS-32QAM separately. We compared the post-FEC BER performances of HPGS/PS/regular-32QAM and PAS-64QAM under the same capacity.

The performances of HPGS/PS-32QAM were shown to be similar, and better than regular-32QAM. What is more, PAS-64QAM could still have 0.2–0.4 dB performance gains over HPGS/PS-32QAM.

For a commercial CCDM-based PS implementation, the power consumption is mainly determined by the block length and the entropy of the CCDM, as well as FEC coding selection. GS-QAM formats will cost more power consumption than regular-QAM formats, because of that a higher complexity DSP circuit is required to be used for GS-QAM-based transceivers to recover the signal from the received data with Gaussian-like distribution. Although the best performance can be theoretically achieved by HPGS-QAM, the extra power consumption arising from PS and GS schemes is nontrivial. Considering that both PS and GS can closely approach the Shannon limit and only limited margin can be reached by HPGS, HPGS may not be in favor by the industry due to the low benefit–cost ratio.

Author Contributions: This research was supervised by I.B.D. and J.A. All works were done by Z.Q.

Conflicts of Interest: The authors declare no conflict of interest.

References

1. Zhalehpour, S.; Lin, J.; Guo, M.; Sepehrian, H.; Zhang, Z.; Rusch, L.A.; Shi, W. All-silicon IQ modulator for 100 GBaud 32QAM transmissions. In Proceedings of the Optical Fiber Communication Conference, San Diego, CA, USA, 3–7 March 2019; p. Th4A.5.
2. Xie, Y.; Geng, Z.; Kong, D.; Zhuang, L.; Lowery, A.J. Selectable-FSR 10-GHz granularity WDM superchannel filter in a reconfigurable photonic integrated circuit. *J. Lightw. Technol.* **2018**, *36*, 2619–2626. [CrossRef]
3. Li, Y.; Yang, M.; Mo, W.; Zhu, S.; Qu, Z.; Djordjevic, I.B.; Kilper, D. Hysteresis-based margin allocation for adaptive coding in SDN-enabled optical networks. In Proceedings of the Optical Fiber Communication Conference (OFC), San Diego, CA, USA, 11–15 March 2018; p. Th1D.2.

4. Li, Y.; Mo, W.; Zhu, S.; Shen, Y.; Yu, J.; Samadi, P.; Bergman, K.; Kilper, D.C. tSDX: Enabling impairment-aware all-optical inter-domain exchange. *J. Lightw. Technol.* **2018**, *36*, 142–154. [CrossRef]
5. Nawazuddin, M.B.S.; Wheeler, N.V.; Hayes, J.R.; Sandoghchi, S.R.; Bradley, T.D.; Jasion, G.T.; Slavík, R.; Richardson, D.J.; Poletti, F. Lotus-shaped negative curvature hollow core fiber with 10.5 dB/km at 1550 nm wavelength. *J. Lightw. Technol.* **2018**, *36*, 1213–1219. [CrossRef]
6. Tamura, Y.; Sakuma, H.; Morita, K.; Suzuki, M.; Yamamoto, Y.; Shimada, K.; Honma, Y.; Sohma, K.; Fujii, T.; Hasegawa, T. The first 0.14-dB/km loss optical fiber and its impact on submarine transmission. *J. Lightw. Technol.* **2018**, *36*, 44–49. [CrossRef]
7. Olsson, S.L.; Eliasson, H.; Astra, E.; Karlsson, M.; Andrekson, P.A. Long-haul optical transmission link using low-noise phase-sensitive amplifiers. *Nat. Commun.* **2018**, *9*, 2513. [CrossRef] [PubMed]
8. Miniscalco, W.J. Erbium-doped glasses for fiber amplifier at 1500 nm. *IEEE J. Lightw. Technol.* **1991**, *9*, 234–250. [CrossRef]
9. Ip, E.; Lau, A.P.T.; Barros, D.J.F.; Kahn, J.M. Coherent detection in optical fiber systems. *Opt. Express* **2008**, *16*, 753–791. [CrossRef]
10. Xu, T.; Jacobsen, G.; Popov, S.; Li, J.; Vanin, E.; Wang, K.; Friberg, A.T.; Zhang, Y. Chromatic dispersion compensation in coherent transmission system using digital filters. *Opt. Express* **2010**, *18*, 16243–16257. [CrossRef]
11. Savory, S.J. Digital filters for coherent optical receivers. *Opt. Express* **2008**, *16*, 804–817. [CrossRef] [PubMed]
12. Qu, Z.; Li, Y.; Mo, W.; Yang, M.; Zhu, S.; Kilper, D.; Djordjevic, I.B. Performance optimization of PM-16QAM transmission system enabled by real-time self-adaptive coding. *Opt. Lett.* **2017**, *42*, 4211–4214. [CrossRef] [PubMed]
13. Huang, M.F.; Tanaka, A.; Ip, E.; Huang, Y.K.; Qian, D.; Zhang, Y.; Zhang, S.; Ji, P.N.; Djordjevic, I.B.; Wang, T.; et al. Terabit/s Nyquist superchannels in high capacity fiber field trials using DP-16QAM and DP-8QAM modulation formats. *J. Lightw. Technol.* **2014**, *32*, 776–782. [CrossRef]
14. Kamalov, V.; Jovanovski, L.; Vusirikala, V.; Zhang, S.; Yaman, F.; Nakamura, K.; Inoue, T.; Mateo, E.; Inada, Y. Evolution from 8QAM live traffic to PS 64-QAM with neural-network based nonlinearity compensation on 11000 km open subsea cable. In Proceedings of the Optical Fiber Communication Conference (OFC), San Diego, CA, USA, 11–15 March 2018; p. Th4D-5.
15. Cai, J.X.; Batshon, H.G.; Mazurczyk, M.; Zhang, H.; Sun, Y.; Sinkin, O.V.; Foursa, D.; Pilipetskii, A.N. 64QAM based coded modulation transmission over transoceanic distance with> 60 Tb/s capacity. In Proceedings of the Optical Fiber Communication Conference, Los Angeles, CA, USA, 22–26 March 2015; p. Th5C-8.
16. Zhang, S.; Yaman, F.; Huang, Y.K.; Downie, J.D.; Zou, D.; Wood, W.A.; Zakharian, A.; Khrapko, R.; Mishra, S.; Nazarov, V.; et al. Capacity-approaching transmission over 6375 km at spectral efficiency of 8.3 bit/s/Hz. In Proceedings of the Optical Fiber Communications Conference and Exhibition (OFC), Anaheim, CA, USA, 20–24 March 2016; pp. 1–3.
17. Cai, J.X.; Batshon, H.G.; Mazurczyk, M.V.; Sinkin, O.V.; Wang, D.; Paskov, M.; Davidson, C.R.; Patterson, W.W.; Turukhin, A.; Bolshtyansky, M.A.; et al. 51.5 Tb/s capacity over 17,107 km in C + L bandwidth using single-mode fibers and nonlinearity compensation. *J. Lightw. Technol.* **2018**, *36*, 2135–2141. [CrossRef]
18. Mo, W.; Zhu, S.; Li, Y.; Kilper, D.C. EDFA wavelength dependent gain spectrum measurement using weak optical probe sampling. *Photon. Technol. Lett.* **2017**, *30*, 177–180. [CrossRef]
19. Winzer, P.J.; Neilson, D.T.; Chraplyvy, A.R. Fiber-optic transmission and networking: The previous 20 and the next 20 years. *Opt. Express* **2018**, *26*, 24190–24239. [CrossRef] [PubMed]
20. Sun, X.; Zou, D.; Qu, Z.; Djordjevic, I.B. Run-time reconfigurable adaptive LDPC coding for optical channels. *Opt. Express* **2018**, *26*, 29319–29329. [CrossRef] [PubMed]
21. Qu, Z. Secure High-Speed Optical Communication Systems. Ph.D. Thesis, The University of Arizona, Tucson, AZ, USA, 2018.
22. Wang, Y.; Okamoto, S.; Kasai, K.; Yoshida, M.; Nakazawa, M. Single-channel 200 Gbit/s, 10 Gsymbol/s-1024 QAM injection-locked coherent transmission over 160 km with a pilot-assisted adaptive equalizer. *Opt. Express* **2018**, *26*, 17015–17024. [CrossRef]
23. Karlsson, M.; Agrell, E. Four-dimensional optimized constellations for coherent optical transmission systems. In Proceedings of the European Conference on Optical Communications (ECOC), Torino, Italy, 19–23 September 2010; pp. 1–6.

24. Qu, Z.; Djordjevic, I.B. Optimal constellation shaping in optical communication systems. In Proceedings of the IEEE International Conference on Transparent Optical Networks (ICTON), Bucharest, Romania, 1–5 July 2018; pp. 1–5.
25. Cho, J.; Chen, X.; Chandrasekhar, S.; Raybon, G.; Dar, R.; Schmalen, L.; Burrows, E.; Adamiecki, A.; Corteselli, S.; Pan, Y.; et al. Trans-atlantic field trial using high spectral efficiency probabilistically shaped 64-QAM and single-carrier real-time 250-Gb/s 16-QAM. *J. Lightw. Technol.* **2018**, *36*, 103–113. [CrossRef]
26. Chen, B.; Okonkwo, C.; Hafermann, H.; Alvarado, A. Increasing achievable information rates via geometric shaping. In Proceedings of the European Conference on Optical Communication (ECOC), Rome, Italy, 23–27 September 2018; pp. 1–3.
27. Qu, Z.; Djordjevic, I.B. FEC Coding for nonuniform QAM. In Proceedings of the Signal Processing in Photonic Communications (SPPCom), New Orleans, LA, USA, 24–27 July 2017; p. SpTu3E.2.
28. Rademacher, G.; Luís, R.S.; Puttnam, B.J.; Eriksson, T.A.; Agrell, E.; Maruyama, R.; Aikawa, K.; Furukawa, H.; Awaji, Y.; Wada, N. 159 Tbit/s C+ L band transmission over 1045 km 3-mode graded-index few-mode fiber. In Proceedings of the Optical Fiber Communication Conference (OFC), San Diego, CA, USA, 11–15 March 2018; p. Th4C-4.
29. Rademacher, G.; Ryf, R.; Fontaine, N.K.; Chen, H.; Essiambre, R.J.; Puttnam, B.J.; Luís, R.S.; Awaji, Y.; Wada, N.; Gross, S.; et al. Long-haul transmission over few-mode fibers with space-division multiplexing. *J. Lightw. Technol.* **2018**, *36*, 1382–1388. [CrossRef]
30. Qu, Z.; Fu, S.; Zhang, M.; Tang, M.; Shum, P.; Liu, D. Analytical investigation on self-homodyne coherent system based on few-mode fiber. *Photon. Technol. Lett.* **2014**, *26*, 74–77. [CrossRef]
31. Igarashi, K.; Soma, D.; Wakayama, Y.; Takeshima, K.; Kawaguchi, Y.; Yoshikane, N.; Tsuritani, T.; Morita, I.; Suzuki, M. Ultra-dense spatial-division-multiplexed optical fiber transmission over 6-mode 19-core fibers. *Opt. Express* **2016**, *24*, 10213–10231. [CrossRef] [PubMed]
32. Krzczanowicz, L.; Iqbal, M.A.; Phillips, I.; Tan, M.; Skvortcov, P.; Harper, P.; Forysiak, W. Low transmission penalty dual-stage broadband discrete Raman amplifier. *Opt. Express* **2018**, *26*, 7091–7097. [CrossRef]
33. Shibahara, K.; Mizuno, T.; Lee, D.; Miyamoto, Y. Advanced MIMO signal processing techniques enabling long-haul dense SDM transmissions. *J. Lightw. Technol.* **2018**, *36*, 336–348. [CrossRef]
34. Fischer, J.K.; Schmidt-Langhorst, C.; Alreesh, S.; Elschner, R.; Frey, F.; Berenguer, P.W.; Molle, L.; Nölle, M.; Schubert, C. Generation, transmission, and detection of 4-D set-partitioning QAM signals. *J. Lightw. Technol.* **2015**, *33*, 1445–1451. [CrossRef]
35. Calderbank, A.R.; Ozarow, L.H. Non-equiprobable signaling on the Gaussian channel. *IEEE Trans. Inf. Theory* **1990**, *36*, 726–740. [CrossRef]
36. Forney, G.D. Trellis shaping. *IEEE Trans. Inf. Theory* **1992**, *38*, 281–300. [CrossRef]
37. Khandani, A.K.; Kabal, P. Shaping multidimensional signal spaces. I. Optimum shaping, shell mapping. *IEEE Trans. Inf. Theory* **1993**, *39*, 1799–1808. [CrossRef]
38. Forney, G.D.; Wei, L.-F. Multidimensional constellations—Part I: Introduction, figures of merit, and generalized cross constellations. *IEEE J. Select. Areas Commun.* **1989**, *7*, 877–892. [CrossRef]
39. Forney, G.D. Multidimensional constellations—Part II: Voronoi constellations. *IEEE J. Select. Areas Commun.* **1989**, *7*, 941–958. [CrossRef]
40. Batshon, H.G.; Djordjevic, I.B.; Xu, L.; Wang, T. Iterative polar quantization-based modulation to achieve channel capacity in ultrahigh-speed optical communication systems. *IEEE Photon. J.* **2010**, *2*, 593–599. [CrossRef]
41. Zhang, S.; Yaman, F. Design and comparison of advanced modulation formats based on generalized mutual information. *J. Lightw. Technol.* **2018**, *36*, 416–423. [CrossRef]
42. Qu, Z.; Djordjevic, I.B. Hybrid probabilistic-geometric shaping in optical communication systems. In Proceedings of the 2018 IEEE Photonics Conference (IPC), Reston, VA, USA, 30 September–4 October 2018; pp. 1–2.
43. Batshon, H.G.; Mazurczyk, M.V.; Cai, J.X.; Sinkin, O.V.; Paskov, M.; Davidson, C.R.; Wang, D.; Bolshtyansky, M.; Foursa, D. Coded modulation based on 56APSK with hybrid shaping for high spectral efficiency transmission. In Proceedings of the 2017 European Conference on Optical Communication (ECOC), Gothenburg, Sweden, 17–21 September 2017; pp. 1–3.

44. Zhang, S.; Qu, Z.; Yaman, F.; Mateo, E.; Inoue, T.; Nakamura, K.; Inada, Y.; Djordjevic, I.B. Flex-rate transmission using hybrid probabilistic and geometric shaped 32QAM. In Proceedings of the Optical Fiber Communication Conference (OFC), San Diego, CA, USA, 11–15 March 2018; p. M1G.3.
45. Liu, T.; Qu, Z.; Lin, C.; Djordjevic, I.B. Non-uniform signaling based LDPC coded modulation for high-speed optical transport networks. In Proceedings of the Asia Communications and Photonics Conference (ACP), Wuhan, China, 2–5 November 2016; p. AF3D.5.
46. Kschischang, F.R.; Pasupathy, S. Optimal nonuniform signaling for Gaussian channels. *IEEE Trans. Inf. Theory* **1993**, *39*, 913–929. [CrossRef]
47. Gallager, R.G. *Information Theory and Reliable Communication*; Wiley: Hoboken, NJ, USA, 1968.
48. Schulte, P.; Bocherer, G. Constant composition distribution matching. *IEEE Trans. Inf. Theory* **2016**, *62*, 430–434. [CrossRef]
49. Fehenberger, T.; Millar, D.S.; Koike-Akino, T.; Kojima, K.; Parsons, K. Multiset-partition distribution matching. *IEEE Trans. Commun.* **2018**, *67*, 1885–1893. [CrossRef]
50. Böcherer, G.; Steiner, F.; Schulte, P. Fast probabilistic shaping implementation for long-haul fiber-optic communication systems. In Proceedings of the 2017 European Conference on Optical Communication (ECOC), Gothenburg, Sweden, 17–21 September 2017; pp. 1–3.
51. Böcherer, G.; Steiner, F.; Schulte, P. Bandwidth efficient and rate-matched low-density parity-check coded modulation. *IEEE Trans. Commun.* **2015**, *63*, 4651–4665. [CrossRef]
52. Fehenberger, T.; Alvarado, A.; Bocherer, G.; Hanik, N. On probabilistic shaping of quadrature amplitude modulation for the nonlinear fiber channel. *J. Lightw. Technol.* **2016**, *34*, 5063–5073. [CrossRef]
53. Ren, J.; Liu, B.; Xu, X.; Zhang, L.; Mao, Y.; Wu, X.; Zhang, Y.; Jiang, L.; Xin, X. A probabilistically shaped star-CAP-16/32 modulation based on constellation design with honeycomb-like decision regions. *Opt. Express* **2019**, *27*, 2732–2746. [CrossRef]
54. Qu, Z.; Djordjevic, I.B. Geometrically shaped 16QAM outperforming probabilistically shaped 16QAM. In Proceedings of the European Conference on Optical Communication (ECOC), Gothenburg, Sweden, 17–21 September 2017; pp. 1–3.
55. Schmalen, L.; Alvarado, A.; Rios-Müller, R. Performance prediction of nonbinary forward error correction in optical transmission experiments. *J. Lightw. Technol.* **2017**, *35*, 1015–1027. [CrossRef]
56. Lin, C.; Qu, Z.; Liu, T.; Zou, D.; Djordjevic, I.B. Experimental study of capacity approaching general LDPC coded non-uniform shaping modulation format. In Proceedings of the Asia Communications and Photonics Conference (ACP), Wuhan, China, 2–5 November 2016; p. AF3A-1.
57. Qu, Z.I.; Djordjevic, I.B. On the probabilistic shaping and geometric shaping in optical communication systems. *IEEE Access* **2019**, *7*, 21454–21464. [CrossRef]
58. Qu, Z.; Lin, C.; Liu, T.; Djordjevic, I.B. Experimental investigation of GF(3(exp 2)) nonbinary LDPC-coded Non-uniform 9-QAM modulation format. In Proceedings of the European Conference on Optical Communication (ECOC), Dusseldorf, Germany, 18–22 September 2016; pp. 1112–1114.
59. Qu, Z.; Lin, C.; Liu, T.; Djordjevic, I.B. Experimental study of nonlinearity tolerant modulation formats based on LDPC coded non-uniform signaling. In Proceedings of the Optical Fiber Communications Conference and Exhibition (OFC), Los Angeles, CA, USA, 19–23 March 2017; p. W1G.7.
60. Qu, Z.; Zhang, S.; Djordjevic, I.B. Universal hybrid probabilistic-geometric shaping based on two-dimensional distribution matchers. In Proceedings of the Optical Fiber Communication Conference (OFC), San Diego, CA, USA, 11–15 March 2018; p. M4E.4.
61. Arnold, D.M.; Loeliger, H.-A.; Vontobel, P.O.; Kavcic, A.; Zeng, W. Simulation-based computation of information rates for channels with memory. *IEEE Trans. Inf. Theory* **2006**, *52*, 3498–3508. [CrossRef]
62. Yankovn, M.P.; Da Ros, F.; da Silva, E.P.; Forchhammer, S.; Larsen, K.J.; Oxenløwe, L.K.; Galili, M.; Zibar, D. Constellation shaping for WDM systems using 256QAM/1024QAM with probabilistic optimization. *J. Lightw. Technol.* **2016**, *34*, 5146–5156. [CrossRef]

© 2019 by the authors. Licensee MDPI, Basel, Switzerland. This article is an open access article distributed under the terms and conditions of the Creative Commons Attribution (CC BY) license (http://creativecommons.org/licenses/by/4.0/).

Article

Joint Probabilistic-Nyquist Pulse Shaping for an LDPC-Coded 8-PAM Signal in DWDM Data Center Communications

Xiao Han [1,2,*], Mingwei Yang [1], Ivan B. Djordjevic [1], Yang Yue [2], Qiang Wang [2], Zhen Qu [2] and Jon Anderson [2]

[1] ECE Department, University of Arizona, Tucson, AZ 85721, USA; mingweiyang@email.arizona.edu (M.Y.); ivan@email.arizona.edu (I.B.D.)
[2] Juniper Networks, 1133 Innovation Way, Sunnyvale, CA 94089, USA; yyue@juniper.net (Y.Y.); qiwang.thresh@gmail.com (Q.W.); zqu@juniper.net (Z.Q.); jonanderson@juniper.net (J.A.)
* Correspondence: xhan322@email.arizona.edu

Received: 1 November 2019; Accepted: 18 November 2019; Published: 20 November 2019

Abstract: M-ary pulse-amplitude modulation (PAM) meets the requirements of data center communication because of its simplicity, but coarse entropy granularity cannot meet the dynamic bandwidth demands, and there is a large capacity gap between uniform formats and the Shannon limit. The dense wavelength division multiplexing (DWDM) system is widely used to increase the channel capacity, but low spectral efficiency of the intensity modulation/direct detection (IM/DD) solution restricts the throughput of the modern DWDM data center networks. Probabilistic shaping distribution is a good candidate to offer us a fine entropy granularity and efficiently reduce the gap to the Shannon limit, and Nyquist pulse shaping is widely used to increase the spectral efficiency. We aim toward the joint usage of probabilistic shaping and Nyquist pulse shaping with low-density parity-check (LDPC) coding to improve the bit error rate (BER) performance of 8-PAM signal transmission. We optimized the code rate of the LDPC code and compared different Nyquist pulse shaping parameters using simulations and experiments. We achieved a 0.43 dB gain using Nyquist pulse shaping, and a 1.1 dB gain using probabilistic shaping, while the joint use of probabilistic shaping and Nyquist pulse shaping achieved a 1.27 dB gain, which offers an excellent improvement without upgrading the transceivers.

Keywords: pulse amplitude modulation; nyquist pulse shaping; DWDM system; LDPC coding

1. Introduction

In view of the current development of the annual growth rate of data center transmission, the widely used coherent optical communication [1–5] is moving toward the data center networks market, but has not dominated because of its high cost, high power consumption, and implementation complexity. Self-coherent detection is a viable solution to reduce the cost, but its expensive coherent receiver still limits its use to a narrow range of applications [6–8]. To match the low-cost requirement, many researchers have paid increasing attention to pulse-amplitude modulation (PAM) [9–11]. However, for the currently widely used uniform distribution, coarse entropy granularity of M-ary PAM (M = 2, 4, 8, ...) cannot meet the dynamic bandwidth demands. More importantly, there exists a large capacity gap between the uniform modulation formats and the Shannon limit.

To compensate for the performance loss, in recent years, a constellation shaping scheme has attracted increasing research attention, which include geometric shaping (GS) [12–17], probabilistic shaping (PS) [18–23], and hybrid geometric-probabilistic shaping [24–27]. These different shaping schemes can approach the Shannon limit. Since it may be easier to have a common standard agreement among network service providers if the PS scheme is used for constellation shaping, PS should be more suitable to be used in data center networks. PS can be realized using Huffman coding [28,29], many-to-one mapping [30], and a constant composition distribution matcher (CCDM) [31,32]. Given that a CCDM can flexibly generate fractional entropy without systematic error, the optical industry is more inclined to apply a CCDM-based PS scheme. PS can not only be used for quadrature amplitude modulation (QAM) formats, but also for a PAM scheme [33–35]. For a PAM scheme, it imposes an exponential-like distribution on a set of equidistant constellation points. It transmits symbols with smaller amplitudes more often than larger ones, which can offer us a fine entropy granularity and enable a transmission with a lower signal-to-noise ratio (SNR) at the same forward error correction (FEC) overhead.

The requirement of channel capacity is greatly increasing nowadays, which leads to the dense wavelength division multiplexing (DWDM) system being widely using in data center networks [36–38], but the low spectral efficiency of the intensity modulation/direct detection (IM/DD) solution restricts the throughput of the modern DWDM data center networks, which results in large investments to upgrade the transceivers in order to meet the increasing bandwidth requirement.

Nyquist pulse shaping (NPS) is a good solution to increase the spectral efficiency of a DWDM system [39,40]; it uses a raised-cosine filter to limit the effective bandwidth and can reduce the inter-symbol interference (ISI) by properly selecting the roll-off factor (ROF) for NPS.

The purpose of this paper was to demonstrate the joint usage of both the PS distribution and NPS in a DWDM system for short-reach applications, in particular data center networks. In addition to the joint shaping format, the employment of a suitable FEC code was also important for improving the overall performance. Low-density parity-check (LDPC) codes represent excellent FEC candidates to be applied together with the proposed shaping scheme. We transmitted LDPC-coded 8-PAM signals for both uniform and nonuniform signaling and compared the performance for different LDPC code rates. We further evaluated the performance improvements when NPS was used for different ROFs in both PS and uniform distribution-based systems. We experimentally evaluated the bit error rate (BER) performance improvement of the proposed joint shaping scheme, compared with the uniform signaling scheme.

The rest of the paper is organized as follows. In Section 2, we introduce the DWDM system employing the proposed joint probabilistic-Nyquist pulse shaping scheme. In Section 3, we demonstrate the improvements with respect to uniform signaling via simulation and experimental verifications. Relevant concluding remarks are provided in Section 4.

2. Proposed PS-NPS-8-PAM-Based DWDM System

Figure 1 shows the proposed LDPC-coded PS-NPS-8-PAM signal generator. The input was a pseudorandom binary sequence (PRBS), and after the distribution matcher (DM), there was an array of different amplitudes that satisfy a certain probability distribution. After binary labelling and LDPC encoding [36], we obtained a block of LDPC-coded binary bits. After mapping to the PAM constellation points, we performed the Nyquist pulse shaping, followed by the DWDM multiplexing.

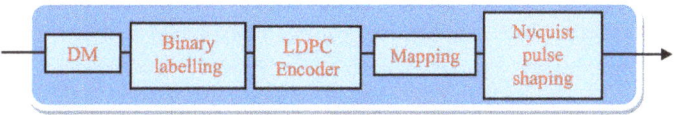

Figure 1. Low-density parity-check (LDPC)-coded, probabilistic shaping, Nyquist pulse shaping, 8-level pulse-amplitude modulation (PS-NPS-8-PAM) signal generator. DM: Distribution matcher.

The PS distribution performances are highly dependent on a selected distribution function. In this paper, we used an exponential distribution [33], in which different constellation points a_i were transmitted with probabilities determined using:

$$P(a_i) = \exp(-\lambda \|a_i\|)/Z(\lambda), \ \lambda \geq 0, \tag{1}$$

where $Z(\lambda)$ is the normalization function used to ensure that the probabilities of occurrence of symbols sum up to one, which means $Z(\lambda)$ is defined as:

$$Z(\lambda) = \sum_i \exp(-\lambda \|a_i\|). \tag{2}$$

Nyquist pulse shaping is widely used in DWDM systems to improve the spectral efficiency [40]. It often uses a low-pass raised-cosine (RC) filter with the frequency response being:

$$H_{RC}(\omega) = \begin{cases} T_s & 0 \leq |\omega| < \pi(1-\text{ROF})/T_S \\ \frac{T_s}{2}\left(1-\sin\left[\frac{T_S}{2\times\text{ROF}}\left(|\omega|-\frac{\pi}{T_S}\right)\right]\right), & \pi(1-\text{ROF})/T_S \leq |\omega| < \pi(1+\text{ROF})/T_S \ , \\ 0 & |\omega| > \pi(1+\text{ROF})/T_S \end{cases} \tag{3}$$

where ω is the angular frequency, T_s is the symbol duration, and ROF is the roll-off factor mentioned before. A small ROF value can make the frequency response an almost rectangular shape and be able to reduce the channel spacing but comes with a longer memory length and a higher generation complexity. In this study, we used the square-root raised cosine (SRRC) filter, whose transfer function is given as:

$$H_{SRRC}(\omega) = \sqrt{H_{RC}(\omega)}. \tag{4}$$

Figure 2 shows the frequency response for two different ROF values of neighboring DWDM channels, with the channel spacing set to 50 GHz. It is clear that a smaller ROF had a better spectral efficiency with a lower inter-channel crosstalk.

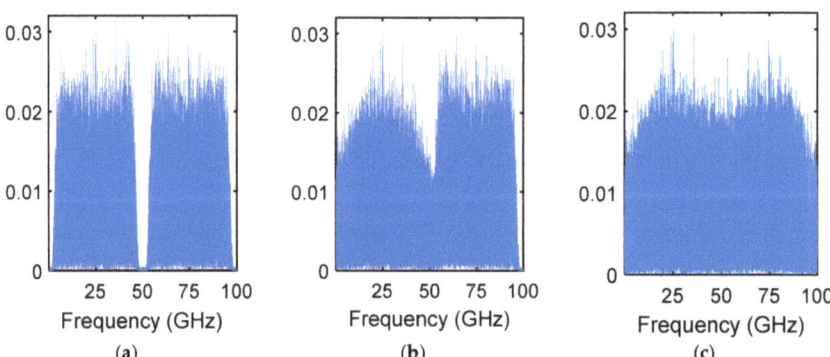

Figure 2. Frequency response with different roll-off factors (ROFs): (**a**) ROF$_1$ = ROF$_2$ = 0.1; (**b**) ROF$_1$ = 1, ROF$_2$ = 0.1; and (**c**) ROF$_1$ = ROF$_2$ = 1.

Figure 3 shows the whole DWDM system. On the transmitter side, we generated the PS-NPS-8-PAM signals using the generator in Figure 1 and sent them to the modulator for each channel with a different frequency. Then, after transmission, at the receiver, we used a super Gaussian filter to select every target frequency (wavelength).

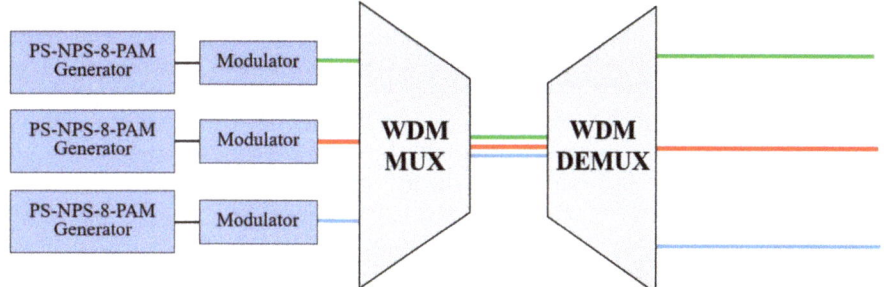

Figure 3. Structure of the PS-NPS-8-PAM-based DWDM system. DEMUX: demultiplexer.

3. Simulation and Data Center Experimental Results

In this section, we first describe the optimization of the code rate of the LDPC encoder for the PS distribution and show a different ROF factor performance using simulations. Then, we introduce the experimental setup and the improvement we achieved from our PS-NPS-8-PAM scheme.

3.1. Simulation Results

Figure 4 shows the BER performance comparison for LDPC-coded 8-PAM signals with PS and uniform distributions for different LDPC code rates. As the PS distribution has a smaller entropy, we needed to make sure that each modulation scheme had the same information rate to guarantee that the comparison was fair; in other words, they had the same FEC overhead. The code rate of the uniform distribution was 0.6, so the information rate was 1.8 bits/symbol, with a 66.7% FEC overhead. From Figure 4, we can see that all PS distributions outperformed the uniform distribution, and the best improvement offered a 0.8 dB signal-to-noise ratio (SNR) gain over the uniform distribution at a BER of 10^{-5}, which appeared when code rate (r) equal to 0.8.

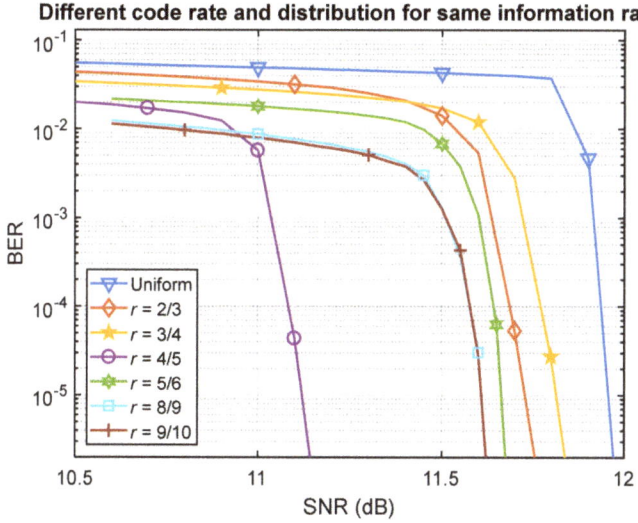

Figure 4. Bit error ratio (BER) performance for different LDPC code rates (r). SNR: Signal-to-noise ratio.

Figure 5 shows the BER performance of the LDPC-coded PS and uniform distribution for different values of the ROF. We can see for both distribution schemes that a smaller ROF achieved a better

BER performance, which was caused by the reason shown in Figure 2: a smaller ROF value can lead to the frequency response of a rectangular shape, improving the spectral efficiency. For every ROF value considered, we found that the PS distribution-based scheme always outperformed uniformly distribution-based one.

Figure 5a shows a very clear error flow when the ROF was set to 0.5. Namely, every LDPC code had an effective SNR region, in which it sufficiently improved the BER performance after a corresponding SNR threshold. According to Figure 2, a larger ROF brought a wider bandwidth and more inter-channel crosstalk, which prevented us reaching the SNR threshold of the employed LDPC code.

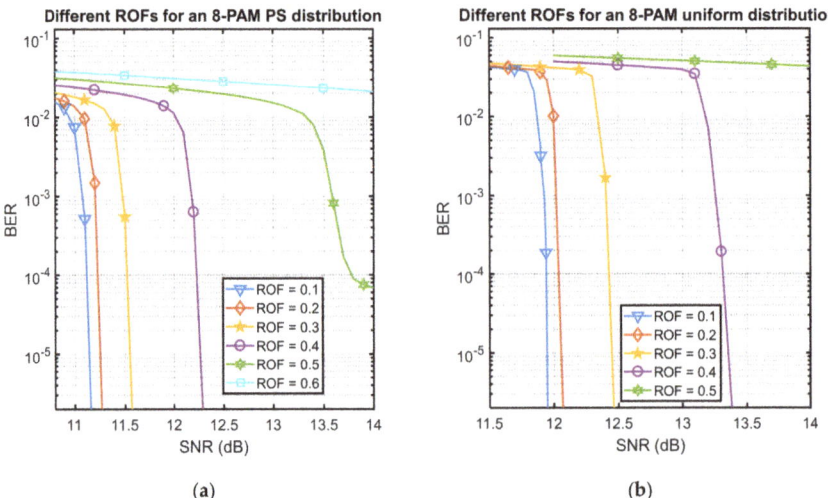

Figure 5. BER performance for different ROF values for: (**a**) a PS distribution with an LDPC code of rate $r = 0.8$ and (**b**) a uniform distribution with an LDPC code of rate $r = 0.6$.

3.2. Experimental Setup and Results

The performance of the LDPC-coded PS-NPS-8-PAM transmission in a fiber channel-based DWDM system was experimentally verified with the testbed depicted in Figure 6. Three 10 kHz-linewidth, continuous-wave, tunable sources (with the following center frequencies: $f_1 = 193.30$ THz, $f_2 = 193.35$ THz, and $f_3 = 193.40$ THz) were coupled by an optical coupler and launched to a Mach-Zehnder modulator. The PRBS was sent to our LDPC-coded PS-NPS-8-PAM generator. Then, the LDPC-coded PS-NPS signals were sent to an arbitrary waveform generator (AWGen) to create 11.5 GBaud 8-PAM signals with a 66.7% FEC overhead, which meant the bit rate was 20.7 Gbps. After being converted to the optical domain by the modulator, the resulting signals were boosted by an erbium-doped fiber amplifier (EDFA) with a 6 dB noise figure. The enhanced signal was then sent to a 1 × 3 coupler to be split and interleaved into three fibers with different lengths (delay lines). The corresponding outputs were then applied to a 32 × 32 arrayed waveguide grating (AWG)-based datacenter network at three different input ports. Every input port worked as a selective bandpass filter. At the output port, we obtained three different center frequencies with different delays, which worked as our DWDM system. The output signal was then mixed with an amplified spontaneous emission (ASE) noise signal using a 2 × 2 coupler. We also employed a variable optical attenuator (VOA) after the ASE noise source to emulate the different optical SNR (OSNR) channel conditions. At the receiver side, the targeted frequency f_2 was selected by a tunable filter (TF) and detected using a photodetector (PD). To collect the received baseband signal, we employed a 100 GS/s digital phosphor oscilloscope from Tektronix. Then, we performed offline digital signal processing (DSP) with the collected signals.

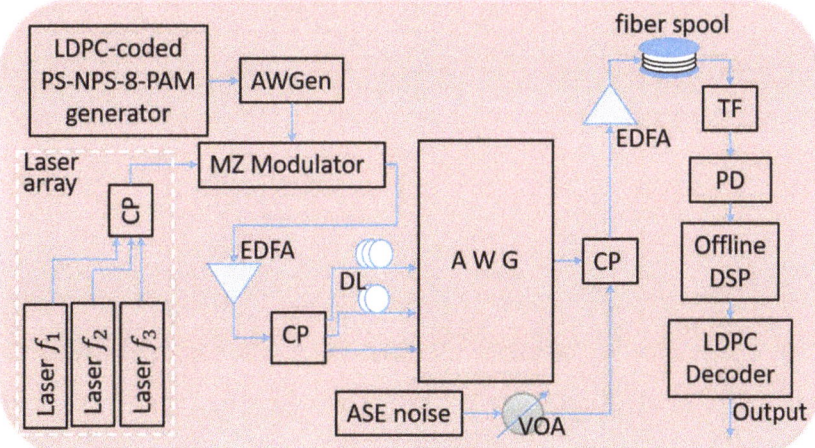

Figure 6. Experimental data center setup. CP: coupler, DL: delay line.

The BER performance of the LDPC-coded PS-NPS-8-PAM is shown in Figure 7a. We transmitted the PS-NPS-8-PAM signals with ROF = 0.8 and 1, and compared them against a PS-8-PAM. From this figure, we can see that when ROF = 1, it performed the same as the PS-8-PAM without NPS, which is because ROF = 1 will not change the pulse shape. On the other hand, when we set the ROF to 0.8, we obtained a 0.43 dB shaping gain improvement in OSNR at BER = 10^{-5}. In Figure 7b, we compare the LDPC-coded PS- and uniform distribution-based schemes with both ROFs set to 0.8. We can see that uniform distribution required higher OSNRs, and we obtained a 1.1 dB shaping gain improvement in OSNR at BER = 10^{-5}.

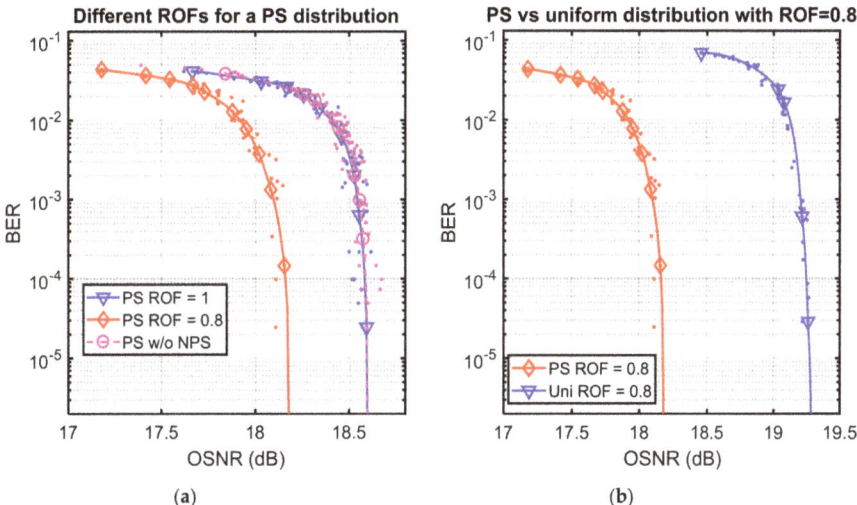

Figure 7. BER performance for: (**a**) a PS distribution with different ROFs and without NPS, and (**b**) a PS and uniform distribution comparison with the same ROF. OSNR: optical signal-to-noise ratio.

Figure 8 shows the comparison of BER performance between the LDPC-coded PS-NPS-8-PAM and LDPC-coded uniform distributed 8-PAM schemes. Evidently, the joint usage of PS and NPS could obtain a 1.27 dB OSNR improvement at BER = 10^{-5} compared to the uniform signaling.

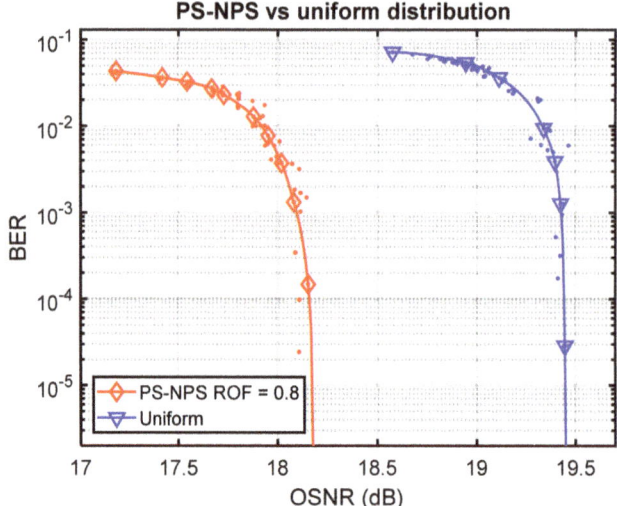

Figure 8. BER performance of an LDPC-coded PS-NPS-8-PAM against an LDPC-coded uniform 8-PAM.

4. Conclusions

In our simulation results, we found that a smaller ROF could provide a better BER performance, and to optimize the PS performance, we needed to consider both the signal entropy and the error correction capability of an LDPC code. In practical usage, there will be many potential limits on the ROF and FEC overhead, since in most cases, we can only use an appropriate ROF for a certain FEC overhead. We have shown that the optimization was not highly sensitive to either the ROF or FEC overhead, and we could do the optimization on any limit of the ROF and FEC overhead. The improvement can be achieved without any equipment upgrade, which means that the proposed LDPC-coded PS-NPS-8-PAM scheme can be widely applied in data center communications and other short-reach applications. We have shown that the Nyquist pulse shaping could provide a 0.43 dB improvement, and the probabilistic shaping provided a 1.1 dB gain. By the joint use of probabilistic shaping and Nyquist pulse shaping, we obtained a 1.27 dB performance gain.

Author Contributions: This research was conducted by X.H. M.Y. offered experimental support. Y.Y., Q.W., Z.Q. and J.A. supervised the NPS and PS idea and simulation during Han's internship at Juniper Networks. Both the simulation and the experimental demonstration were supervised by I.B.D.

Funding: This research was funded by Juniper Networks and NSF.

Conflicts of Interest: The authors declare no conflict of interest.

References

1. Huang, M.F.; Tanaka, A.; Ip, E.; Huang, Y.K.; Qian, D.; Zhang, Y.; Zhang, S.; Ji, P.N.; Djordjevic, I.B.; Wang, T.; et al. Terabit/s Nyquist superchannels in high capacity fiber field trials using DP-16QAM and DP-8QAM modulation formats. *J. Lightwave Technol.* **2014**, *32*, 776–782. [CrossRef]
2. Perin, J.K.; Shastri, A.; Kahn, J.M. Design of Low-Power DSP-Free Coherent Receivers for Data Center Links. *J. Lightwave Technol.* **2017**, *35*, 4650–4662. [CrossRef]
3. Cai, J.X.; Batshon, H.G.; Mazurczyk, M.; Zhang, H.; Sun, Y.; Sinkin, O.V.; Foursa, D.; Pilipetskii, A.N. 64QAM based coded modulation transmission over transoceanic distance with > 60Tb/s capacity. In Proceedings of the Optical Fiber Communication Conference (OFC), Los Angeles, CA, USA, 22–26 March 2015.
4. Qu, Z.; Li, Y.; Mo, W.; Yang, M.; Zhu, S.; Kilper, D.; Djordjevic, I.B. Performance optimization of PM-16QAM transmission system enabled by real-time self-adaptive coding. *Opt. Lett.* **2017**, *42*, 4211–4214. [CrossRef]

5. Winzer, P.J.; Neilson, D.T.; Chraplyvy, A.R. Fiber-optic transmission and networking: The previous 20 and the next 20 years. *Opt. Express* **2018**, *26*, 24190–24239. [CrossRef]
6. Puttnam, B.J.; Luis, R.; Delgado-Mendinueta, J.-M.; Sakaguchi, J.; Klaus, W.; Awaji, Y.; Wada, N.; Kanno, A.; Kawanishi, T. High-capacity self-homodyne PDM-WDM-SDM transmission in a 19-core fiber. *Opt. Express* **2014**, *22*, 21185–21191. [CrossRef]
7. Qu, Z.; Fu, S.; Zhang, M.; Tang, M.; Shum, P.; Liu, D. Analytical investigation on self-homodyne coherent system based on few-mode fiber. *IEEE Photonics Technol. Lett.* **2014**, *26*, 74–77. [CrossRef]
8. Puttnam, B.J.; Luís, R.S.; Mendinueta, J.M.D.; Sakaguchi, J.; Klaus, W.; Kamio, Y.; Nakamura, M.; Wada, N.; Awaji, Y.; Kanno, A.; et al. Self-homodyne detection in optical communication systems. *Photonics* **2014**, *1*, 110–130. [CrossRef]
9. Zhong, K.; Zhou, X.; Gao, Y.; Chen, W.; Man, J.; Zeng, L.; Lau, A.P.T.; Lu, C. 140-Gb/s 20-km Transmission of PAM-4 Signal at 1.3 µm for Short Reach Communications. *IEEE Photonics Technol. Lett.* **2015**, *27*, 1757–1760. [CrossRef]
10. Eiselt, N.; Wei, J.; Griesser, H.; Dochhan, A.; Eiselt, M.; Elbers, J.-P.; Olmos, J.J.V.; Monroy, I.T. First real-time 400G PAM-4 demonstration for inter-data center transmission over 100 km of SSMF at 1550 nm. In Proceedings of the Optical Fiber Communications Conference and Exhibition (OFC), Anaheim, CA, USA, 20–24 March 2016.
11. Mestre, M.A.; Mardoyan, H.; Konczykowska, A.; Rios-Müller, R.; Renaudier, J.; Jorge, F.; Duval, B.; Dupuy, J.-Y.; Ghazisaeidi, A.; Jennevé, P.; et al. Direct detection transceiver at 150-Gbit/s net data rate using PAM 8 for optical interconnects. In Proceedings of the 2015 European Conference on Optical Communication (ECOC), Valencia, Spain, 27 September–1 October 2015.
12. Chen, B.; Okonkwo, C.; Hafermann, H.; Alvarado, A. Increasing achievable information rates via geometric shaping. In Proceedings of the European Conference on Optical Communication (ECOC), Rome, Italy, 23–27 September 2018.
13. Qu, Z.; Djordjevic, I.B.; Anderson, J. Two-Dimensional Constellation Shaping in Fiber-Optic Communications. *Appl. Sci.* **2019**, *9*, 1889. [CrossRef]
14. Khandani, A.K.; Kabal, P. Shaping multidimensional signal spaces. I. Optimum shaping, shell mapping. *IEEE Trans. Inf. Theory* **1993**, *39*, 1799–1808. [CrossRef]
15. Qu, Z.; Djordjevic, I.B. Geometrically shaped 16QAM outperforming probabilistically shaped 16QAM. In Proceedings of the European Conference on Optical Communication (ECOC), Gothenburg, Sweden, 17–21 September 2017.
16. Forney, G.D. Trellis shaping. *IEEE Trans. Inf. Theory* **1992**, *38*, 281–300. [CrossRef]
17. Forney, G.D.; Wei, L.-F. Multidimensional constellations—Part I: Introduction, figures of merit, and generalized cross constellations. *IEEE J. Sel. Areas Commun.* **1989**, *7*, 877–892. [CrossRef]
18. Cho, J.; Chen, X.; Chandrasekhar, S.; Raybon, G.; Dar, R.; Schmalen, L.; Burrows, E.; Adamiecki, A.; Cortesseli, S.; Pan, Y.; et al. Trans-atlantic field trial using high spectral efficiency probabilistically shaped 64-QAM and single-carrier real-time 250-Gb/s 16-QAM. *J. Lightware Technol.* **2018**, *36*, 103–113. [CrossRef]
19. Calderbank, A.R.; Ozarow, L.H. Non-equiprobable signaling on the Gaussian channel. *IEEE Trans. Inf. Theory* **1990**, *36*, 726–740. [CrossRef]
20. Qu, Z.; Djordjevic, I.B. On the Probabilistic Shaping and Geometric Shaping in Optical Communication Systems. *IEEE Access* **2019**, *7*, 21454–21464. [CrossRef]
21. Steiner, F.; Böcherer, G. Comparison of Geometric and Probabilistic Shaping with Application to ATSC 3.0. In Proceedings of the International ITG Conference on Systems, Communications and Coding (SCC), Hamburg, Germany, 6–9 February 2017.
22. Fehenberger, T.; Alvarado, A.; Bocherer, G.; Hanik, N. On probabilistic shaping of quadrature amplitude modulation for the nonlinear fiber channel. *J. Lightware Technol.* **2016**, *34*, 5063–5073. [CrossRef]
23. Böcherer, G.; Steiner, F.; Schulte, P. Bandwidth efficient and rate-matched low-density parity-check coded modulation. *IEEE Trans. Commun.* **2015**, *63*, 4651–4665. [CrossRef]
24. Qu, Z.; Djordjevic, I.B. Hybrid Probabilistic-Geometric Shaping in Optical Communication Systems. In Proceedings of the IEEE Photonics Conference (IPC), Reston, VA, USA, 30 September–4 October 2018.

25. Batshon, H.G.; Mazurczyk, M.V.; Cai, J.X.; Sinkin, O.V.; Paskov, M.; Davidson, C.R.; Wang, D.; Bolshtyansky, M.; Foursa, D. Coded modulation based on 56APSK with hybrid shaping for high spectral efficiency transmission. In Proceedings of the 2017 European Conference on Optical Communication (ECOC), Gothenburg, Sweden, 17–21 September 2017.
26. Ren, J.; Liu, B.; Xu, X.; Zhang, L.; Mao, Y.; Wu, X.; Zhang, Y.; Jiang, L.; Xin, X. A probabilistically shaped star-CAP-16/32 modulation based on constellation design with honeycomb-like decision regions. *Opt. Express* **2019**, *27*, 2732–2746. [CrossRef]
27. Han, X.; Yang, M.; Djordjevic, I.B. Hybrid Probabilistic-Geometric-Shaped 8-PAM Suitable for Data Centers' Communication. In Proceedings of the Asia Communications and Photonics Conference (ACP), Hangzhou, China, 26–29 October 2018.
28. Qu, Z.; Lin, C.; Liu, T.; Djordjevic, I.B. Experimental study of nonlinearity tolerant modulation formats based on LDPC coded non-uniform signaling. In Proceedings of the Optical Fiber Communications Conference and Exhibition (OFC), Los Angeles, CA, USA, 19–23 March 2017.
29. Yankovn, M.P.; Ros, F.D.; Silva, E.P.; Forchhammer, S.; Larsen, K.J.; Oxenløwe, L.K.; Galili, M.; Zibar, D. Constellation Shaping for WDM Systems Using 256QAM/1024QAM with Probabilistic Optimization. *J. Lightware Technol.* **2016**, *34*, 5146–5156. [CrossRef]
30. Gallager, R.G. *Information Theory and Reliable Communication*; Wiley: Hoboken, NJ, USA, 1968.
31. Schulte, P.; Bocherer, G. Constant composition distribution matching. *IEEE Trans. Inf. Theory* **2016**, *62*, 430–434. [CrossRef]
32. Fehenberger, T.; Millar, D.S.; Koike-Akino, T.; Kojima, K.; Parsons, K. Multiset-partition distribution matching. *IEEE Trans. Commun.* **2018**, *67*, 1885–1893. [CrossRef]
33. Han, X.; Djordjevic, I.B. Probabilistically shaped 8-PAM suitable for data centers communication. In Proceedings of the International Conference on Transparent Optical Networks (ICTON), Bucharest, Romania, 1–5 July 2018.
34. Eriksson, T.A.; Chagnon, M.; Buchali, F.; Schuh, K.; Brink, S.T.; Schmalen, L. 56 Gbaud Probabilistically Shaped PAM8 for Data Center Interconnects. In Proceedings of the 2017 European Conference on Optical Communication (ECOC), Gothenburg, Sweden, 17–21 September 2017.
35. Sun, L.; Wang, C.; Du, J.; Liang, C.; Zhang, W.; Xu, K.; Zhang, F.; He, Z. Dyadic Probabilistic Shaping of PAM-4 and PAM-8 for Cost-Effective VCSEL-MMF Optical Interconnection. *IEEE Photonics J.* **2019**, *11*, 1–12. [CrossRef]
36. Lu, P.; Zhang, L.; Liu, X.; Yao, J.; Zhu, Z. Highly efficient data migration and backup for big data applications in elastic optical inter-data-center networks. *IEEE Netw.* **2015**, *29*, 36–42. [CrossRef]
37. Mahimkar, A.; Chiu, A.; Doverspike, R.; Feuer, M.D.; Magill, P.; Mavrogiorgis, E.; Pastor, J.; Woodward, S.L.; Yates, J. Bandwidth on Demand for Inter-Data Center Communication. In Proceedings of the 10th ACM Workshop on Hot Topics in Networks (HotNets-X'11), New York, NY, USA, 14–15 November 2011; ACM: New York, NY, USA, 2011.
38. Chen, K.; Singla, A.; Singh, A.; Ramachandran, K.; Xu, L.; Zhang, Y.; Wen, X.; Chen, Y. OSA: An Optical Switching Architecture for Data Center Networks with Unprecedented Flexibility. *IEEE/ACM Trans. Netw.* **2014**, *22*, 498–511. [CrossRef]
39. Hirai, R.; Kikuchi, N.; Fukui, T. High-spectral efficiency DWDM transmission of 100-Gbit/s/lambda IM/DD single sideband-baseband-Nyquist-PAM8 signals. In Proceedings of the Optical Fiber Communications Conference and Exhibition (OFC), Los Angeles, CA, USA, 19–23 March 2017.
40. Yue, Y.; Wang, Q.; Anderson, J. Transmitter skew tolerance and spectral efficiency tradeoff in high baud-rate QAM optical communication systems. *Opt. Express* **2018**, *26*, 15045–15058. [CrossRef]

 © 2019 by the authors. Licensee MDPI, Basel, Switzerland. This article is an open access article distributed under the terms and conditions of the Creative Commons Attribution (CC BY) license (http://creativecommons.org/licenses/by/4.0/).

Article

Adaptive Compensation of Bandwidth Narrowing Effect for Coherent In-Phase Quadrature Transponder through Finite Impulse Response Filter

Qiang Wang, Yang Yue *, Jian Yao and Jon Anderson

Juniper Networks, 1133 Innovation Way, Sunnyvale, CA 94089, USA; qiwang.thresh@gmail.com (Q.W.); jianyao@juniper.net (J.Y.); joanderson@juniper.net (J.A.)
* Correspondence: yyue@juniper.net; Tel.: +14-087-452-000

Received: 5 February 2019; Accepted: 8 May 2019; Published: 13 May 2019

Abstract: Coherent in-phase quadrature (IQ) transponders are ubiquitous in the long-haul and the metro optical networks. During the transmission, the coherent signal experiences a bandwidth narrowing effect after passing through multiple reconfigurable optical add-drop multiplexers (ROADMs). The coherent signal also experiences a bandwidth narrowing effect when electrical or optical components of the coherent IQ transponder experience aging. A dynamic method to compensate the bandwidth narrowing effect is thus required. In the coherent optical receiver, signal bandwidth is estimated from the raw analog-to-digital converter (ADC) outputs. By adaptively adjusting the tap coefficients of the finite impulse response (FIR) filter, simple post-ADC FIR filters can increase the resiliency of the coherent signal to the bandwidth narrowing effect. The influence of chromatic dispersion, polarization mode dispersion, and polarization dependent loss are studied comprehensively. Furthermore, the bandwidth information of the transmitted analog signal is fed back to the coherent optical transmitter for signal optimization, and the transmitter-side FIR filter thus changes accordingly.

Keywords: coherent communications; optical communications; fiber optics; digital signal processing

1. Introduction

Coherent line-cards using the polarization-division-multiplexed quadrature amplitude modulation (PDM-QAM) have been the de-facto standards for the metro and the long-haul optical fiber communications systems [1]. When a modulated signal passes through multiple reconfigurable optical add-drop multiplexers (ROADMs), the bandwidth of the propagating signal narrows down due to two effects. One is the spectral shape of the ROADM filter, and the other is the misalignment between the central frequency of the signal and the ROADM passband [2,3]. This inevitably leads to an increase in the observed bit error ratio (BER) at the receiving end and eventually causes the receiver to lose track of the signal. Moreover, the bandwidth narrowing effect due to the ROADM is relatively dynamic.

Several methods have been studied to quantify the effect of cascading ROADMs. In [4], an accurate model is developed to predict the final bandwidth of the signal after passing through multiple cascaded ROADMs. In [5,6], the weighted crosstalk method is used to estimate the transmission penalty, including the effects of signal bandwidth narrowing, crosstalk filtering, and nonlinear-enhance crosstalk. In [7], the impact of the random group delay ripple on multiple cascading ROADMs is thoroughly studied. In [8], one observes signal-to-noise ratio (SNR) for *n* times and determines the minimal SNR. Then, the probability of this minimal SNR being smaller than a specified target is evaluated through extreme value statistics. This approach improves the reliability to access the performance of cascaded ROADMs. In [9], the influence of the bandwidth narrowing effect on a probabilistically shaped 64-QAM system is evaluated using generalized mutual information (GMI). The shaping parameters can be optimized to improve the tolerance to the bandwidth narrowing effect.

With a comprehensive understanding from the theoretical models and experimental results above, different schemes to compensate the effects of the bandwidth narrowing have been demonstrated. In [10], digital pre-equalization is applied to the coherent transmitter based on the estimation of the signal's bandwidth using an analytical model. The parameters in the analytical model are determined by the statistics from the measurement results. In [11], the maximum likelihood sequence estimation (MLSE) is used to mitigate inter-symbol interference (ISI) induced by the bandwidth narrowing effect. In [12], time-domain digital pre-equalization is used for a bandwidth-limited coherent communications system. The characteristics of the channel are generated from the coherent receiver through a digital adaptive equalizer. In [13], the bandwidth narrowing effect is studied in the context of elastic optical network. Different techniques, like bandwidth-variable transmitter and optical spectrum shaping, are used to compensate the bandwidth narrowing effect. In [14], the bandwidth-variable transceivers offer flexibility in adjusting both the symbol rate and the spectral efficiency. With the cascaded ROADMs, the balance between the optical filtering tolerance and the required SNR can be optimized. In [15], the data stream from the transmitter passes through a digital filter and generates the duobinary data pattern, which is more tolerant to the bandwidth narrowing effect.

In this work, we propose a novel scheme to compensate the bandwidth narrowing effect by simply using the finite impulse response (FIR) filters. First, we demonstrate that the bandwidth of the received signal can be estimated from the raw analog-to-digital converter (ADC) data samples. Then, we adaptively adjust the FIR filter in the coherent receiver to provide compensation according to the received signal bandwidth. The tolerance of the coherent receiver to the bandwidth narrowing effect is improved by 5 GHz, which is equivalent to the bandwidth degradation of the signal passing through >10 ROADMs [16]. Furthermore, the estimated signal bandwidth can be fed back to the coherent transmitter. By adaptively adjusting the FIR filter in the coherent transmitter, one can dynamically change the spectral shape of the coherent signal. The system tolerance to the bandwidth narrowing effect through the cascaded ROADMs is further improved. This scheme is relatively simple to implement, and it is also complementary to the other methods discussed above [10–15].

2. Experimental Setup and Signal Bandwidth Estimation

On top of the optical transport network, there is an Internet protocol (IP) network utilizing the packet forwarding engine (PFE) application-specific integrated circuit (ASIC) to forward the IP packets. It is very desirable to physically integrate the coherent transponder with the PFE, which will remove any client-side optical transponders to reduce the cost and power consumption. Figure 1 shows the architecture, which closely integrates the PFE, the digital signal processing (DSP) ASIC, and the coherent optical transponders. The client's signal is received by the DSP and converted to the polarization-division-multiplexed quadrature phase shift keying (PDM-QPSK) modulation format at 30.1 giga-baud (GBd). The coherent optical transponder is a highly integrated C form factor two analog coherent optics (CFP2-ACO) module [17]. The payload of each transponder is either 100 gigabit Ethernet (GbE) or Optical Channel Transport Unit 4 (OTU4).

The IP traffic is converted into optical signal through the coherent transmitter and then transmitted through the long-haul optical communications system, which can have multiple cascading ROADMs. During the transmission, multiple optical impairments, such as chromatic dispersion (CD), polarization mode dispersion (PMD), and polarization dependent loss (PDL), can accumulate. At the receiver, the optical signal is coherently detected by beating with the local oscillator (LO). Then, the signal is converted back to the digital domain through the ADC. Most of the optical impairments are compensated by the DSP ASIC. Figure 1 depicts this detailed configuration. In addition, a network management layer oversees the whole long-haul system.

The digital analog converter (DAC), the circuit trace, the connector for pluggable optics, the radio-frequency (RF) amplifier, and the electro-optical modulator form the analog interface between the DSP chip and the coherent transmitter (Tx). Together with the ADC, the photodiode, the trans-impedance amplifier, the circuit trace, and the connector in the pluggable ACO form the analog interface between the coherent receiver (Rx) and the DSP chip. The received signal bandwidth is influenced by the bandwidth of both these analog interfaces. Although this influence is relatively static, it might degrade over the lifetime of the coherent transponder. Thus, a dynamic mechanism to compensate the bandwidth narrowing effect would also improve the tolerance to the component degradation.

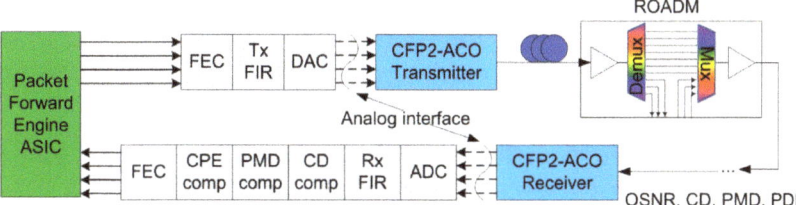

Figure 1. Block diagram of coherent line-card and optical line system. FEC: forward error correction, CPE: carrier phase estimation. Network management layer is not shown for simplicity.

Figure 2 shows the experimental setup to study the bandwidth narrowing effect. CD, PMD, and PDL emulators are used to emulate various impairments in the transmission system. The 4 × 4 non-blocking optical switches can reconfigurably load either individual impairment or combined impairments. The noise produced by the amplified spontaneous emission (ASE) is added to adjust the optical signal-to-noise ratio (OSNR). An optical filter whose center frequency and pass-band are tunable is placed in front of the coherent receiver. The tunable filter is Finisar's WaveShaper built on the liquid crystal on silicon (LCOS) technology. The filter's passband closely resembles that of a wavelength selective switch (WSS), which is widely used in modern ROADM [4]. One noticeable difference is that with multiple cascading ROADMs, there is a steep roll-off at high-frequency content of the filter's frequency response, which cannot be emulated by the WaveShaper. Despite this slight difference, one can adjust the pass-band of the tunable filter, allowing the emulation of the bandwidth narrowing effect to the coherent signal.

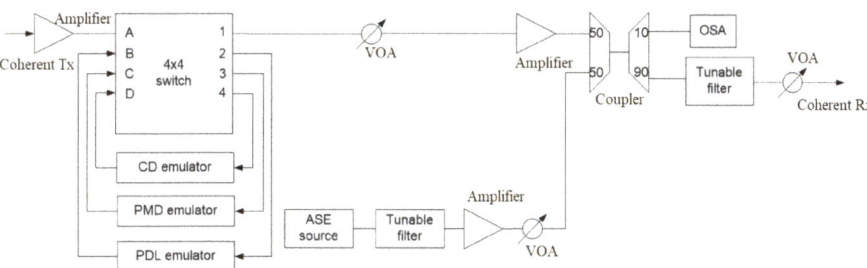

Figure 2. Experimental setup to emulate the bandwidth narrowing effect together with other impairments during the transmission. VOA: variable optical attenuator, OSA: optical spectrum analyzer.

To achieve adaptive compensation of the bandwidth narrowing effect, we firstly estimate the signal's bandwidth. The coherent DSP ASIC can provide a snapshot of the ADC raw data and store them in the random-access memory (RAM). The network management layer can extract the ADC raw data and perform fast Fourier transform (FFT) to calculate the spectrum of the modulated signal. The sample rate of the ADC is 45 GHz. There are 16,384 points for the ADC raw data, and FFT is then applied to the whole data set. The raw spectra from FFT are shown in Figure 3a. However, the data modulation makes it difficult to estimate the exact bandwidth. Thus, we apply the Savitzky-Golay filter [18] to smooth the raw spectrum. The length of the filter tap is 11, and the polynomial order of the filter is three. The Savitzky-Golay filter can improve the SNR with minimum distortion to the signal. It performs much better than the averaging filter, which filters out a significant portion of high frequency content of the signal. Figure 3b shows the spectrum after the Savitzky-Golay filter being applied.

Next, we adjust the bandwidth of the optical tunable filter to emulate the narrowing effect for the coherent signal's bandwidth. We define this parameter as the signal's bandwidth. The spectra of the X-polarization in-phase tributary (XI) under different bandwidth are shown in Figure 3c. Clearly, the bandwidth narrowing effect is presented in the spectrum of the received signal. We defined the difference between the signal's power at the spectrum's edge (Nyquist frequency) and the one at the spectrum's center as $Attn_{Edge}$. This parameter is closely related to the spectral shape of the signal and can be used to characterize the bandwidth of the signal. This value was further averaged over four tributaries to improve the accuracy. Figure 4 shows a roughly linear relationship between $Attn_{Edge}$ and the signal's bandwidth. There is small variation for $Attn_{Edge}$ over different measurements, thus multiple measurements are used to improve the accuracy.

It is also noticeable that $Attn_{Edge}$ is dependent on the SNR level. To estimate the signal's bandwidth (BW) from $Attn_{Edge}$, one must approximately determine the SNR. In our experiment, the noise is emulated by an external ASE source. In such a condition, the SNR is approximately equivalent to the OSNR. One can thus use the pre-FEC BER to estimate the SNR [19].

In the long-haul optical communications system utilizing dense wavelength division multiplexing (DWDM), the noise is also coming from the nonlinear effect. In a modern DWDM system where coherent transponder is widely deployed, CD is continuously accumulated, and it is not compensated by the inline dispersion compensation module (DCM) anymore. Consequently, uncompensated CD broadens the spectrum of a coherent signal to a Gaussian-like shape. In the nonlinear regime where the per-channel power is larger than the optimal launching power, cross phase modulation (XPM) is the main source of fiber nonlinearity. However, most systems operate in the linear regime or weakly nonlinear regime where the per-channel power is smaller than the optimal launching power. Here, four wave mixing (FWM) is the main source of fiber nonlinearity. The noise generated from FWM approximately acts as the additive white Gaussian noise (AWGN) to the channel under consideration due to its Gaussian-like spectrum [20]. As a result, SNR is further reduced to a value smaller than OSNR. A well-known Gaussian noise (GN) model has been developed as a simple yet reliable tool to predict the performance of uncompensated coherent systems [21]. Thus, the pre-FEC BER can also be used to estimate SNR using the GN model.

Based on the discussion above, the measured $Attn_{Edge}$ can be used to reversely estimate the signal's bandwidth at the coherent receiver. In turn, the FIR filter in the receiver's data path can be adaptive adjusted to compensate the bandwidth-narrowing effect based on the value of $Attn_{Edge}$.

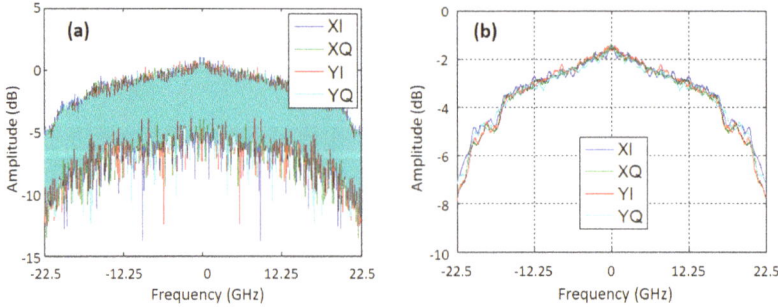

Figure 3. Cont.

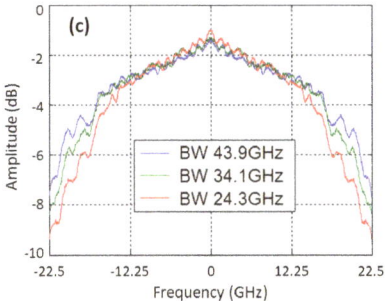

Figure 3. (**a**) Spectra from the analog-to-digital converter (ADC)'s raw outputs. (**b**) Spectra after applying the Savitzky-Golay filter. (**c**) Spectra of XI tributary under different bandwidths. Results are measured with 14-dB optical signal-to-noise ratio (OSNR).

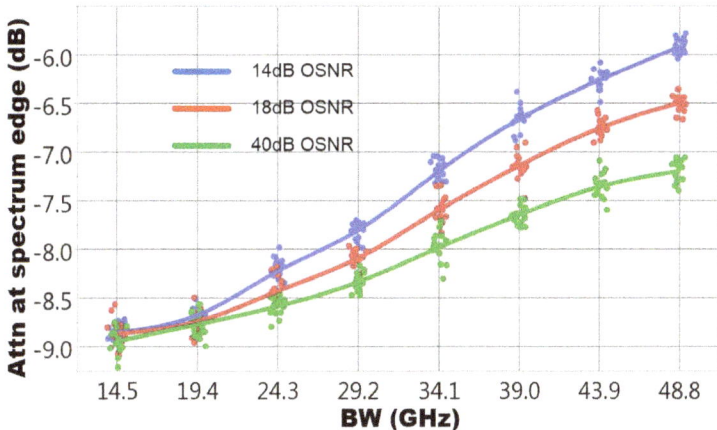

Figure 4. Attenuation at spectrum edge ($Attn_{Edge}$) versus the signal's bandwidth in the receiver. For each setting of signal's bandwidth, we measure 16 times and take the average value.

3. Adaptive Compensation through Digital Filters

Simple FIR filters can be placed right after the ADC and before the frequency-domain CD compensation module. The typical usage of those simple FIR filters is to compensate the RF losses of the analog interface, and the tap values of the FIR filters usually remain unchanged during the lifetime of the coherent transponder. However, in this work, the tap coefficients of the FIR filters are dynamically optimized, particularly to compensate the bandwidth narrowing effect. As discussed in Section 2, the value of $Attn_{Edge}$ is directly correlated to the signal's bandwidth. One can use this value to see whether the signal's bandwidth is limited. If so, one can adjust the simple FIR filter to provide more peaking to compensate the limited bandwidth.

Two key parameters determine the shape of channel filters—peaking frequency and peaking amount. The peaking amount is the difference between the maximal frequency response and the frequency response close to the direct current (DC). The peaking frequency is where the maximal response locates in the frequency domain. The output of the FIR filter is a convolution between the input signal and the FIR's impulse response, as shown in the equation below.

$$y(n) = \sum_{j=1}^{N} h(j)x(n-j) \qquad (1)$$

Here, x is the input signal to the FIR filter, h is the tap coefficient of the FIR filter, and N is the total number of taps. In the current DSP ASIC, N is equal to three. The FIR filter is $T_s/2$ spaced, where T_s is the symbol period. The tap coefficients of those digital filters are real numbers only. Also, four sets of the FIR filters (for four tributaries) share the same tap coefficients. These simplifications greatly reduce the complexity in the hardware implementation.

Next, we search different combinations of the tap coefficients and identify the desired spectral shapes. The desired range for the peaking frequency is from 13 GHz to 19 GHz, and the one for the peaking amount is from 1 dB to 5 dB. To obtain the target spectral shape, we first set the value of main tap to be one. Next, we perform a coarse scan on the values of precursor and postcursor. We adjust those values between −0.5 to 0.5 at the step of 0.1. By performing a Z-transform, the frequency response of FIR filter is determined. Then, we identify a small range of precursor and postcursor. Within this small range, the peaking amount and the peaking frequency are close to the target values. Furthermore, we optimize the values of precursor and postcursor by performing a fine scan within this small range. With only three taps, we sometimes cannot obtain both the desired peaking amount and the peaking frequency simultaneously. In those situations, we prioritize the peaking amount over the peaking frequency. Finally, we convert the representation of tap coefficients from floating-point number to signed-integer number. The conversion is based on the following rules—the tap coefficients are implemented by 8-bits registers with the most significant bit (MSB) being the sign bit; for the remaining seven bits, two bits are used to represent integer value, and five bits are used to represent fractional value. Figure 5 shows spectral shape dependence of a few FIR filters on the peaking frequency and the peaking amount. One advantage of those simple FIR filters is their approximately linear phase response, which minimizes the ripple in the frequency domain.

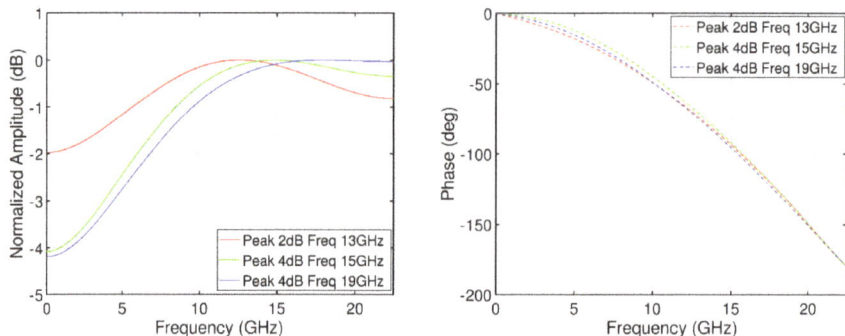

Figure 5. Normalized amplitude and phase responses of simple finite impulse response (FIR) filters with different peaking frequencies and peaking amounts. The results are obtained by applying Z-transform on the tap coefficients of the FIR filter.

For each filter shape, there are multiple sets of tap coefficients satisfying the requirement. It is also important to consider the DC gain of the FIR filter, which is determined by the summation of the absolute value of those tap coefficients. A FIR filter with a high DC gain will lead the signal to being clipped, while a FIR filter with a low DC gain will lead the signal to being submerged under the noise. We optimize the DC gain of the FIR filter to achieve the best performance. We choose three typical filters: 2-dB peaking at 13 GHz, 4-dB peaking at 15 GHz, and 4-dB peaking at 19 GHz, corresponding to the three filters shown in Figure 5. We adjust the DC gain of the FIR filter while maintaining the spectral shape in the frequency domain. We measure the BER under different DC gain and summarize the results in Figure 6. As seen, for each filter's shape, there is an optimal DC gain setting leading to minimum BER. In the following sections, we set the DC gain of the FIR filter to the optimal point.

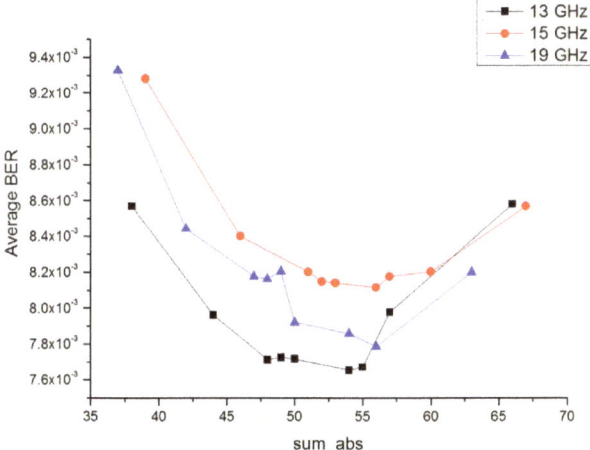

Figure 6. The direct current (DC) gain of the FIR filter should be optimized to minimize the bit error ratio (BER).

In general, the spectral response of the FIR filter should be inverse to the overall frequency response of the cascaded ROADMs for equalization. When a signal has a relatively high bandwidth with a small number of cascading ROADMs, the optimal filter has a small peaking amplitude and a small peaking frequency. Otherwise, more noise will be amplified. When a signal has a low bandwidth due to larger number of cascading ROADMs, the optimal filter should have a large peaking amplitude and a large peaking frequency. In this way, any reduction of the signal's bandwidth will be compensated by this adaptive FIR filter.

We first study the scenario where no FIR filter is applied. However, we obtain a subpar BER result. The performance degradation is much more severe when various impairments are included. The reason is that the optical and the electrical components in the receiver's data path accumulate the limitation on the received signal's bandwidth. Even when the receiver's signal bandwidth is not limited by the filter-narrowing effect due to the cascading ROADMs, a FIR filter with certain peaking amount is still needed to compensate the bandwidth limitation introduced by various components in the receiver's data path, as discussed in the introduction. We experimentally identify that a mild FIR filter with 2-dB peaking at 13 GHz provides the optimal results across different scenarios. The performance is similar to that reported in [22], where a coherent transponder is built with discrete premium components. This demonstrates that the mild FIR compensates the bandwidth limitation of integrated coherent receiver.

To demonstrate the advantage of our proposed method, we also select a strong FIR filter with 4-dB peaking at 19 GHz for further study. It is expected that the mild FIR filter will perform better when the signal's bandwidth is large, while the strong FIR filter will perform better when the signal's bandwidth is small. Through the experimental measurements under different transmission impairments, we validate this prediction. The results are summarized in Figure 7.

For these two settings, we measure the pre-FEC BER at 16-dB OSNR with a receiver optical power of −18 dBm. This is close to our system's end-of-life specification, which is the summation of the required OSNR for the coherent receiver and the reserved margin for operation. The pre-FEC BER is reported by the DSP ASIC using the output of the FEC decoder. Furthermore, one can use the pre-FEC BER to estimate the SNR. We first convert the measured BER to the Q^2 factor. Next, we choose the baseline as the measured Q^2 factor in the back-to-back scenario with the strong FIR setting. We determine the penalty by calculating the difference between the Q^2 factor under the specific scenario and the Q^2 factor of the baseline scenario.

In the subplot of the back-to-back scenario, the penalty of the Q^2 factor under the strong FIR setting is always zero, since this is chosen as the baseline. With the signal's bandwidth larger than 40 GHz, we obtain a negative penalty for the mild setting, which indicates a better BER performance. However, when the signal's bandwidth is smaller than 30 GHz, we obtain a positive penalty for the mild setting, which indicates a worse BER performance. In the other scenarios, the penalty in the Q^2 factor is always positive due to the fact that the transmission impairments such as CD, PMD, and PDL cause the degradation in the BER. Still, we can notice that the mild FIR setting performs better with the signal's bandwidth larger than 40 GHz, while the strong FIR setting performs better with the signal's bandwidth smaller than 30 GHz. Those experimental results agree with the analysis above. In the scenarios of back-to-back, 3-dB PDL, and combined impairment, as shown in Figure 7a–c, a coherent receiver using a strong FIR setting can still recover the signal with 20-GHz signal's bandwidth; however, a coherent receiver using the mild FIR setting fails to recover this signal. This indicates that the operation range of the coherent receiver using 4-dB peaking at 19 GHz (strong FIR setting) is 5 GHz larger than that of the coherent receiver using 2-dB peaking at 13 GHz (mild FIR setting). A 5-GHz improvement on the tolerance to the bandwidth narrowing effect will allow the coherent signal to pass 10 additional ROADMs [16].

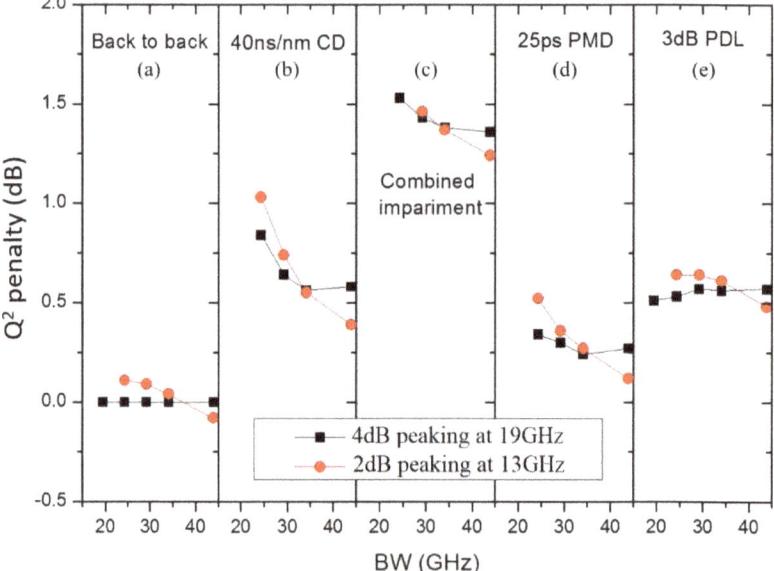

Figure 7. The penalty in Q^2 factor versus the receiver bandwidth under different scenarios of the transmission impairments. (**a**) Result in back-to-back setup. (**b**) Result with 40 ps/nm CD. (**c**) Result with combined impairment. (**d**) Result with 25 ps PMD. (**e**) Result with 3 dB PDL. The Combined impairments are 3-dB polarization dependent loss (PDL), 18-ps polarization mode dispersion (PMD), and 40-ns/nm chromatic dispersion (CD).

In the data path of the coherent receiver, there is another adaptive equalizer (AEQ) to track the state of polarization (SoP), de-multiplex the orthogonal polarizations, and compensate the PMD. It is illustrated as the "PMD Comp" block in Figure 1. The AEQ also has multiple taps, thus a large amount of PMD can be compensated. To some degree, the bandwidth-narrowing effect can be partially compensated by this AEQ. The advantages of AEQ are high loop bandwidth and small response time. Thus, the equalizer can quickly track and dynamically compensate the bandwidth narrowing effect. However, there are certain limitations associated with the AEQ. The first limitation from AEQ is its complexity. The AEQ is a 2 × 2 equalizer with tap coefficients being complex numbers. It has large power consumption and long latency as well. The second limitation is that AEQ is placed after the CD compensation block. Thus, the bandwidth narrowing effect will increase the noise floor of the CD compensation block. The third limitation is that the quality of the signal impacts whether AEQ converges. When the bandwidth of the signal is severely narrowed down to <20 GHz, we notice that the coherent receiver cannot lock due to the fact that the AEQ cannot converge anymore. This can be seen from Figure 7a,c,e, where there are no data points for the mild FIR setting. When we apply the strong FIR setting to partially compensate the bandwidth narrowing effect, the AEQ converges, and the coherent receiver locks. We summarize the difference between two types of equalizers in Table 1.

Table 1. Comparison between Rx FIR filter and PMD adaptive equalizer.

Parameters	Rx FIR Filter	PMD AEQ (Adaptive Equalizer)
Main function	Static loss compensation for RF channel	SoP tracking and PMD compensation
Location	Before CD block	After CD block
Tap coefficients	Real value	Complex value
Power consumption	Small	Large
Complexity	Simple tap-and-delay	Complicate 2 × 2 butterfly
Feedback mechanism	Spectrum from ADC raw data	Blind equalization or data-aided equalization
Loop bandwidth	Low	High
Number of taps	Small	Moderate
Impact of signal's bandwidth	Minimal	AEQ may not converge with small bandwidth

The influence of these two types of equalizers on the optical performance of the coherent transponder can also be viewed in the experimental results above. For example, in Figure 7d, where the impairment of 25-ps PMD is presented, we notice that the penalty in Q^2 factor for the mild FIR setting is increased from 0.1 dB to 0.5 dB when the signal's bandwidth is reduced from 50 GHz to 20 GHz. This shows that, even though the AEQ for PMD compensation can adjust its spectral response, the impairment due to the signal's bandwidth narrowing effect cannot be fully mitigated by the AEQ alone. By changing the simple Rx FIR filter from the mild setting (2-dB peaking at 13 GHz) to the strong setting (4-dB peaking at 19 GHz), the penalty in Q^2 factor is reduced from 0.5 dB to 0.3 dB with 20 GHz signal's bandwidth, demonstrating the performance improvement from the simple Rx FIR filter on top of the AEQ. Similarly, this performance improvement can also be observed when other types of optical impairments are introduced, as shown in Figure 7. Overall, both the AEQ for PMD compensation and the Rx FIR for channel loss compensation can be simultaneously used to improve the tolerance to the bandwidth narrowing effect of the coherent signal passing through multiple cascading ROADMs. A complementary method is to use the PMD equalizer to compensate the dynamic change of the signal's bandwidth and use the simple FIR filter to compensate the slow drift of the signal's bandwidth.

In addition to post-compensation at the Rx side, it is beneficial to apply pre-compensation on the Tx side as well. The Tx FIR is located after the FEC layer and before the DAC input. Most applications use those FIR filters to compensate the relatively static loss of the RF channel between the DAC's output and the Mach-Zehnder modulator (MZM) in a set-and-forget approach. Since we can estimate the BW of the signal, we can also adaptively adjust the Tx FIR to pre-compensate the narrowing effect of signal's bandwidth. In addition, the Nyquist filter is widely used to improve the spectral efficiency, which is usually a raised cosine (RC) filter or a root raised cosine (RRC) filter. The roll-off factor for the RC filter or the RRC filter determines the spectral response [23]. Thus, the key parameters of the FIR in the coherent transmitter are the peaking amplitude, the peaking frequency, and the roll-off factor. To determine the tap coefficients, the frequency response of the FIR filter with the desired peaking frequency and the peaking amount is first multiplied with the frequency response of the Nyquist filter. Next, a reverse Z-transform is performed to solve the coefficients of the combined filter.

As a demonstration, we apply the peaking filter and the Nyquist filter together to the Tx FIR. The number of taps in the Tx FIR is 17, and the FIR filter is $T_s/2$ spaced. We use the RRC filter with the roll-off factor of one as the Nyquist filter. We adjust the peaking amount from 1 dB to 9 dB while keeping the peaking frequency at the Nyquist frequency. We measure the optical spectra using a high-resolution optical spectrum analyzer (Finisar Wave Analyzer 1500) and summarize the results in Figure 8. As seen, the optical spectra clearly show the influence of different peaking values, demonstrating the feasibility of using Tx FIR to pre-compensate the bandwidth narrowing effect.

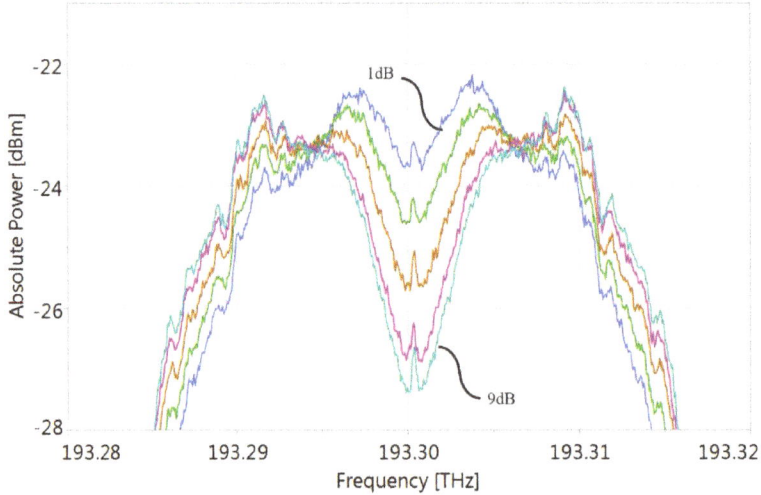

Figure 8. Tx pre-compensation is achieved through the adjustment of FIR filter. The spectra above show the output from the coherent transmitter at different peaking values from 1 dB to 9 dB at 2 dB step size.

The capability of the FIR filter to adjust the signal's spectrum is limited by the total number of taps. Usually, the FIR filters in the Tx have many taps and are independently controlled for each tributary. Thus, the Tx FIR allows great capability and flexibility for pre-compensation of the bandwidth narrowing effect. In general, with the bandwidth narrowing effect, one should increase the peaking amplitude, increase the peaking frequency, and reduce the roll-off factor. These parameters can be adjusted independently or simultaneously.

We summarize the experimental parameters in Table 2 below for both the Tx FIR filter and the Rx FIR filter.

Table 2. Experimental parameters for the Rx FIR filter and the Tx FIR filter.

Parameters	Value and Range
Receiver Optical Power (dBm)	−18
Transmitter Optical Power (dBm)	1
Optical Signal to Noise Ratio (dB)	16
Baud Rate (GBd)	30.1
Chromatic Dispersion (ns/nm)	40
Polarization Mode Dispersion (ps)	25
Polarization Dependent Loss (dB)	3
Signal Bandwidth (GHz)	20 to 50
Rx FIR Peaking Frequency (GHz)	13 to 19
Rx FIR Peaking Amount (dB)	1 to 5
Tx FIR Peaking Amount (dB)	1 to 9

4. Application Scenarios for Different Network Topologies

Different application scenarios exist with various network topologies. For each network topology, there is a different approach to apply the demonstrated method, as depicted in Figure 9. For most of the scenarios in the network, there are bi-directional links. Here, the transmitter of the network node one on the east end goes to the receiver network node two on the west end, while the transmitter of the network node two on the west end goes to the receiver of the network node one on the east end. The links in both directions go through the same number of ROADMs. The ROADMs in both directions are usually of the same type, from the same vendor, and under the same ambient environment. Thus, we can assume that the bandwidth narrowing effect in the east–west direction is the same as that in the west–east direction. In this case, one can estimate the bandwidth from the Rx of node 2 on the west end and use this information to adjust the FIR filter on the Tx of node 2. Hence, the characteristics of the west–east link are improved. This approach is shown in Figure 9a.

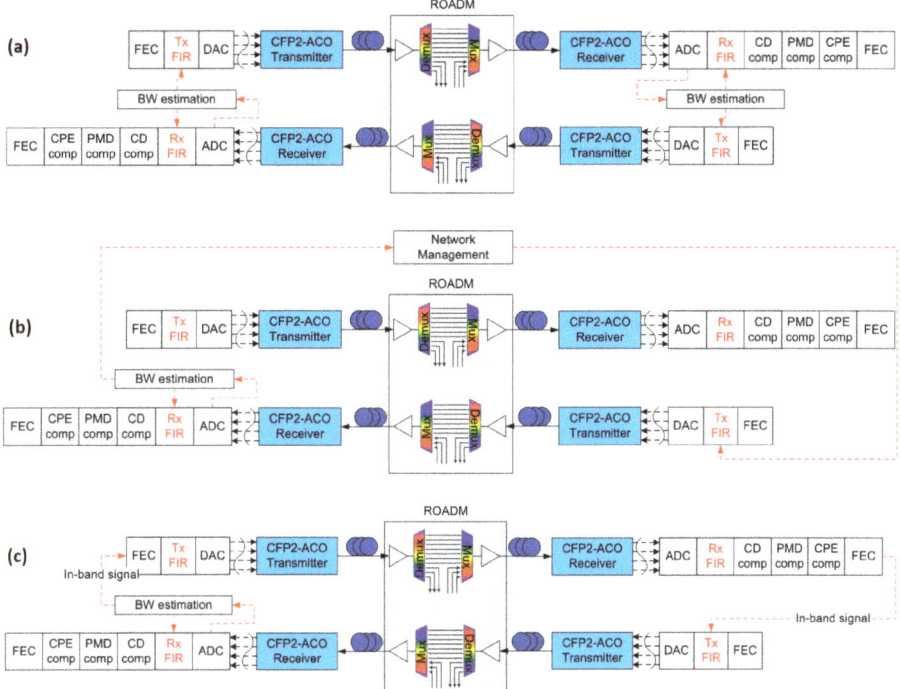

Figure 9. Application scenarios for different network topologies. (**a**) The estimated signal bandwidth is directly fed back from the coherent receiver to the coherent transmitter in a bi-directional link. (**b**) The estimated signal bandwidth is fed back through the network management layer. (**c**) The estimated signal bandwidth is fed back through the in-band signal.

Furthermore, one can also send the estimated information of the signal's bandwidth to the network management layer. The network layer will relay the information to the other end of the link accordingly. This approach is shown in Figure 9b. In addition, another approach is to send the information of the bandwidth in the reverse direction through the in-band signal, which can be embedded in Internet protocol (IP) packet or optical transport network (OTN) frame. This approach is shown in Figure 9c. Once the information is received by the other end, the FIR filter on the Tx side can be configured accordingly. For example, the Rx side of node two estimates the signal bandwidth in the east–west direction. This information is sent through the Tx of node two to the Rx of node one in the west–east direction through two approaches above. Once the node one receives the bandwidth information, it can adaptively adjust its Tx FIR for the pre-compensation that improves the performance of the east–west link. These two approaches can extend the application to any network scenario, although it will take longer time to adjust the Tx FIR to the appropriate setting.

A typical process is depicted below. Initially the FIR filters at both the Rx and the Tx sides are loaded with the default tap coefficients to compensate the channel loss of the analog interfaces. Next, the bandwidth of the received signal is estimated, and the Rx FIR is adaptively adjusted. Since one can calibrate the optical spectrum of the Tx signal and estimate the electrical spectrum of the Rx signal based on the raw data from the ADC, one can derive the degradation due to multiple cascading ROADMs. This information can be then fed back to the Tx directly in the bi-directional case or through the in-band signal/network management layer in the other case. Then, one can adjust the Tx FIR filter in one step based on a certain pre-calibrated table. Next, the Rx FIR filter can be adaptively adjusted to compensate any dynamic bandwidth narrowing effect.

The adaptive nature of the demonstrated method is well suited for real deployment. In the case with cascaded ROADMs, the spectral response of multiple filters will vary inadvertently. By estimating the spectrum of the received signal from the raw data of the ADC samples, one can dynamically adjust the simple FIR filter, which is implemented after the ADC, to compensate the bandwidth narrowing effect. This leads to an improved performance for the long-haul optical communications system.

5. Conclusions

Both the pre-compensation from the transmitter's FIR filter and the post-compensation from the receiver's FIR filter are demonstrated to compensate the bandwidth narrowing effect. The bandwidth of the received coherent signal is estimated from the raw output of the ADC with the improved accuracy by using Savitzky-Golay digital filter. No coherent receiver locking or data recovery is needed. The adaptive equalization of the FIR filter can improve the system's tolerance to the bandwidth narrowing effect. This process can be done in real time with running traffic. The proposed method is critical for the coherent signal with high baud rate and high spectral efficiency, such as a single-carrier 64-GBd 16-QAM signal running with 400-Gb/s information bits.

6. Patents

A US patent was granted for the work presented in this paper.

Author Contributions: Conceptualization, Q.W. and Y.Y.; methodology, Q.W.; software, Q.W. and Y.Y.; validation, Y.Y.; formal analysis, Q.W., Y.Y. and J.Y.; investigation, Y.Y., Q.W., J.Y. and J.A.; resources, Y.Y., Q.W.; writing—original draft preparation, Q.W.; writing—review and editing, Q.W., Y.Y., J.Y. and J.A.; visualization, Q.W., Y.Y., J.Y.; supervision, J.A.

Funding: This research received no external funding.

Acknowledgments: The authors gratefully acknowledge the vigorous encouragement and strong support for innovation from Domenico Di Mola at Juniper Networks.

Conflicts of Interest: The authors declare no conflict of interest.

References

1. Roberts, K.; Zhuge, Q.; Monga, I.; Gareau, S.; Laperle, C. Beyond 100 Gb/s: Capacity, Flexibility, and Network Optimization. *J. Opt. Commun. Netw.* **2017**, *4*, C12–C24. [CrossRef]
2. Tibuleac, S.; Filer, M. Transmission Impairments in DWDM Networks with Reconfigurable Optical Add-Drop Multiplexers. *J. Lightwave Technol.* **2010**, *28*, 557–568. [CrossRef]
3. Rafique, D.; Ellis, A. Nonlinear and ROADM Induced Penalties in 28 Gbaud Dynamic Optical Mesh Networks Employing Electronic Signal Processing. *Opt. Express* **2011**, *19*, 16739–16748. [CrossRef] [PubMed]
4. Pulikkaseril, C.; Stewart, L.A.; Roelens, M.A.; Baxter, G.W.; Poole, S.; Frisken, S. Spectral Modeling of Channel Band Shapes in Wavelength Selective Switches. *Opt. Express* **2011**, *19*, 8458–8470. [CrossRef] [PubMed]
5. Filer, M.; Tibuleac, S. Generalized Weighted Crosstalk for DWDM Systems with Cascaded Wavelength-Selective Switches. *Opt. Express* **2012**, *20*, 17620–17631. [CrossRef] [PubMed]

6. Hsueh, Y.; Stark, A.; Liu, C.; Detwiler, T.; Tibuleac, S.; Filer, M.; Chang, G.; Ralph, S. Passband Narrowing and Crosstalk Impairments in ROADM-Enabled 100G DWDM Networks. *J. Lightwave Technol.* **2012**, *30*, 3980–3986. [CrossRef]
7. Duill, S.; Barry, L. Impact of Random Group Delay Ripple on the Cascadability of Wavelength Select Switches for Nyquist-pulse-shaped Signals. *J. Lightwave Technol.* **2016**, *34*, 4161–4167. [CrossRef]
8. Cartledge, J.; Matos, F.; Laperle, C.; Borowiec, A.; O'Sullivan, M.; Roberts, K. Use of Extreme Value Statistics to Assess the Performance Implications of Cascaded ROADMs. *J. Lightwave Technol.* **2017**, *35*, 5208–5214. [CrossRef]
9. Li, L.; El-Rahman, A.; Cartledge, J. Effect of Bandwidth Narrowing due to Cascaded Wavelength Selective Switches on the Generalized Mutual Information of Probabilistically Shaped 64-QAM Systems. In Proceedings of the 2018 European Conference on Optical Communication (ECOC), Rome, Italy, 23–25 September 2018.
10. Pan, J.; Tibuleac, S. Real-Time ROADM Filtering Penalty Characterization and Generalized Precompensation for Flexible Grid Networks. *IEEE Photonics J.* **2017**, *9*, 7202210. [CrossRef]
11. Jia, Z.; Cai, Y.; Chien, H.; Yu, J. Performance Comparison of Spectrum-narrowing Equalizations with Maximum Likelihood Sequence Estimation and Soft-decision Output. *Opt. Express* **2014**, *22*, 6047–6059. [CrossRef] [PubMed]
12. Zhang, J.; Yu, J.; Chi, N.; Chien, H. Time-domain Digital Pre-equalization for Band Limited Signals Based on Receiver-side Adaptive Equalizers. *Opt. Express* **2014**, *22*, 20515–20529. [CrossRef] [PubMed]
13. Fabrega, J.; Moreolo, M.; Martín, L.; Piat, A.; Riccardi, E.; Roccato, D.; Sambo, N.; Cugini, F.; Poti, L.; Yan, S.; et al. On the Filter Narrowing Issues in Elastic Optical Networks. *J. Opt. Commun. Netw.* **2016**, *8*, A23–A33. [CrossRef]
14. Zhou, X.; Zhuge, Q.; Qiu, M.; Xiang, M.; Zhang, F.; Wu, B.; Qiu, K.; Plant, D. On the Capacity Improvement Achieved by Bandwidth-variable Transceivers in Meshed Optical Networks with Cascaded ROADMs. *Opt. Express* **2017**, *25*, 4773–4782. [CrossRef] [PubMed]
15. Hu, Q.; Buchali, F.; Chagnon, M.; Schuh, K.; Bülow, H. 3.6-Tbps Duobinary 16-QAM Transmission with Improved Tolerance to Cascaded ROADM Filtering Penalty. In Proceedings of the 2018 European Conference on Optical Communication (ECOC), Rome, Italy, 23–25 September 2018.
16. Filer, M.; Tibuleac, S. N-degree ROADM Architecture Comparison: Broadcast-and-Select versus Route-and-Select in 120 Gb/s DP-QPSK Transmission Systems. In Proceedings of the 2014 Optical Fiber Communications Conference and Exposition (OFC), San Francisco, CA, USA, 9–14 March 2014.
17. Lu, F.; Zhang, B.; Yue, Y.; Anderson, J.; Chang, G. Investigation of Pre-Equalization Technique for Pluggable CFP2-ACO Transceivers in Beyond 100 Gb/s Transmissions. *J. Lightwave Technol.* **2017**, *2*, 230–237. [CrossRef]
18. Savitzky, A.; Golay, M. Smoothing and Differentiation of Data by Simplified Least Squares Procedures. *Anal. Chem.* **1964**, *8*, 1627–1639. [CrossRef]
19. Wang, Q.; Yue, Y.; He, X.; Vovan, A.; Anderson, J. Accurate Model to Predict Performance of Coherent Optical Transponder for High Baud Rate and Advanced Modulation Format. *Opt. Express* **2018**, *26*, 12970–12984. [CrossRef] [PubMed]
20. Stark, A.; Hsueh, Y.; Detwiler, T.; Filer, M.; Tibuleac, S.; Ralph, S. System Performance Prediction with the Gaussian Noise Model in 100G PDM-QPSK Coherent Optical Networks. *J. Lightwave Technol.* **2013**, *21*, 3352–3360. [CrossRef]
21. Poggiolini, P.; Bosco, G.; Carena, A.; Curri, V.; Jiang, Y.; Forghieri, F. The GN-Model of Fiber Non-Linear Propagation and its Applications. *J. Lightwave Technol.* **2014**, *32*, 694–721. [CrossRef]
22. Birk, M.; Gerard, P.; Curto, R.; Nelson, L.; Zhou, X.; Magill, P.; Schmidt, T.; Malouin, C.; Zhang, B.; Ibragimov, E.; et al. Real-Time Single-Carrier Coherent 100 Gb/s PM-QPSK Field Trial. *J. Lightwave Technol.* **2011**, *29*, 417–425. [CrossRef]
23. Yue, Y.; Wang, Q.; Anderson, J. Transmitter skew tolerance and spectral efficiency tradeoff in high baud-rate QAM optical communication systems. *Opt. Express* **2018**, *26*, 15045–15058. [CrossRef] [PubMed]

© 2019 by the authors. Licensee MDPI, Basel, Switzerland. This article is an open access article distributed under the terms and conditions of the Creative Commons Attribution (CC BY) license (http://creativecommons.org/licenses/by/4.0/).

Article

A Novel Coding Based Dimming Scheme with Constant Transmission Efficiency in VLC Systems

Yu Zuo and Jian Zhang *

National Digital Switching System Engineering & Technological Research Center, Zhengzhou 450000, China; 3120100548@zju.edu.cn
* Correspondence: Zhang_xinda@126.com

Received: 24 January 2019; Accepted: 22 February 2019; Published: 25 February 2019

Abstract: Visible light communication (VLC) has attracted tremendous attention due to two functions: communication and illumination. Both reliable data transmission and lighting quality need to be considered when the transmitted signal is designed. To achieve the desired levels of illumination, dimming control is an essential technology applied in VLC systems. In this paper, we propose a block coding-based dimming scheme to construct the codeword set, where dimming control can be achieved by changing the ratio of two levels (ON and OFF) based on on-off keying (OOK) modulation. Simulation results show that the proposed scheme can maintain good error performance with constant transmission efficiency under various dimming levels.

Keywords: VLC; block code; dimming control; encoding/decoding algorithm

1. Introduction

Recently, light-emitting diode (LED) lighting technology has made significant progress in indoor applications of green lighting [1,2]. Utilizing the fast response characteristic of LEDs, visible light communication (VLC) systems can simultaneously provide high quality illumination and high-speed wireless data transmission [3,4]. As a complementary technique to radio frequency (RF), VLC has several advantages, e.g., huge bandwidth, high rate transmission, excellent security, and immunity to electromagnetic interference [5]. There is no doubt that VLC is becoming an increasingly attractive communication option. Meanwhile, dimming control plays an essential role in indoor VLC systems, where users can maintain the variable dimming levels. Due to ever-increasing energy consumption, the interest in dimming control has further increased. However, dimming control may cause adverse effects on communication. To overcome these challenges, some dimming schemes have been proposed in recent years [6–12].

Generally, the existing on-off keying (OOK)-based dimming schemes change the ratio of the two levels (ON and OFF) to achieve dimming control. Thus, the average light intensity can be changed to maintain the various brightness demands. In recent years, many works have been carried out to achieve dimming control. The analog dimming scheme [7] adjusts the direct current (DC) bias of the transmitted signal to meet the required dimming levels where light intensity is reduced proportionally to the current. Unfortunately, this approach may change the wavelength of the emitted light, causing a chromaticity shift effect. The variable-OOK (VOOK) scheme proposed in [7] is the combination of OOK and the pulse width modulation (PWM) signal. The inactive portions of the duty cycle are filled by the filler bits with either ones or zeros according to the dimming level. Variable pulse position modulation (VPPM), where 2-PPM is combined with the PWM signal, is responsible for dimming and data transmission. The multiple-PPM (MPPM)-based dimming scheme constructs the codeword set to achieve the theoretical transmission efficiency limit as the codeword length increases [8]. However, the above-mentioned dimming schemes encounter the common problem that the transmission efficiency cannot be fixed under different dimming levels, which may cause a severe effect on data transmission.

In order to solve the problem of the transmission efficiency not being constant under different dimming levels, we propose a block coding-based dimming scheme to achieve data transmission and dimming control simultaneously. Via bitwise AND operation of the block code and dimming code, new codewords are generated to satisfy the dimming requirement. Simulation results show that the proposed scheme can maintain constant transmission efficiency under different dimming levels, and it proves an attractive alternative to the dimming scheme.

The remainder of this paper is organized as follows: In Section 2, the system model of the proposed block coding scheme is given. In Section 3, the implementation of the proposed coding scheme is demonstrated and the construction of the encoding/decoding algorithm is summarized accordingly. In Section 4, simulations are carried out to evaluate the error performance of the proposed scheme. Finally, we conclude our findings in Section 5.

2. System Model

In this section, we first introduce the system model of the proposed scheme. For simplification, intensity modulation and direct detection (IM/DD) can be applied in the VLC system, where the message is carried by the light intensity emitted from the LED. In general, LEDs can be modeled as Lambertian emitters, and the line of sight (LOS) path is considered for the indoor VLC system where reflected light is much weaker then direct light [13]. At the receiver side, we assume perfect symbol synchronization. After the optical channel, the signal detected by the receiver can be expressed as:

$$\mathbf{y} = \mathbf{H}\mathbf{x} + \mathbf{n}, \tag{1}$$

where $\mathbf{x}$ denotes the transmitted optical signal and $\mathbf{n}$ can be considered as additive white Gaussian noise (AWGN) with variance σ^2. H represents the optical channel gain in VLC links, which is given as [1]:

$$H = \begin{cases} \frac{\rho A_r (m+1)}{2\pi D^2} \cos^m(\phi) T_s(\psi) g(\psi) \cos(\psi), & 0 \leq \psi \leq \Psi_C \\ 0, & \psi > \Psi_C \end{cases} \tag{2}$$

As shown in Figure 1, m denotes the order of the Lambertian emission, defined as $m = \frac{-\ln 2}{\ln(\cos \Phi_{1/2})}$. ρ is the receiver responsivity. A_r denotes the effective receiver area of the receiver, and D is the distance between the transceivers. ϕ and ψ are the angle of irradiance and incidence from the LED to the Photodetector (PD), respectively. $T_s(\psi)$ is the gain of the optical filter, and $g(\psi)$ is the gain of the optical concentrator. Ψ_C denotes the field of view (FOV) of the receiver.

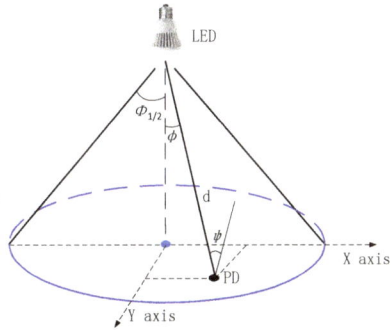

Figure 1. Optical channel model.

Unlike RF systems, the transmitted optical signal should meet the practical lighting constraints, which should be restricted to be nonnegative and less than the maximum limit of light intensity. Without loss of generality, we make the peak light intensity $P = 1$. Meanwhile, considering the dimming function, the average light intensity of the transmitted optical signal is defined as $\mathbb{E}(\mathbf{x}) = \gamma P$, with dimming level $\gamma \in (0, 1)$. For reference, the whole process of the proposed dimming scheme is shown in Figure 2.

Dimming controller: According to the dimming level γ, the "dimming controller" puts parameter K into the "message generation" part and the "dimming code" part, generating block code $\mathbf{b}$ and dimming code $\mathbf{g}$, respectively.

Dimming encoding: With the input block code and dimming code, the "dimming encoding" makes the two parts a bitwise AND operation to generate the transmitted codewords $\mathbf{c}$ satisfying the dimming requirement.

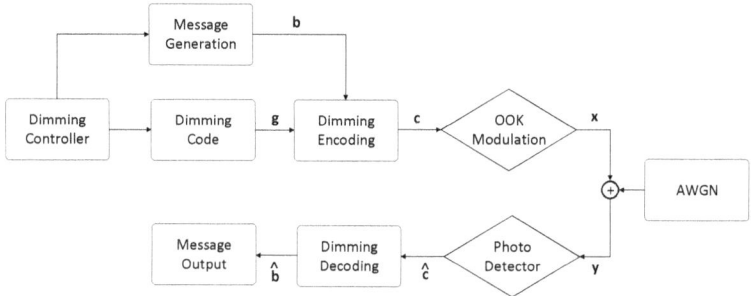

Figure 2. Block diagram of the proposed scheme. OOK, on-off keying.

3. The Proposed Coding Scheme

In VLC links, the brightness can be determined by the average light intensity of the signal. We can achieve dimming control by adjusting the average intensity of the optical signal by constructing the transmitted codeword set. In this section, we will demonstrate the process of the encoding and decoding structure and give the algorithms accordingly.

3.1. Dimming Encoder

In this subsection, the encoding structure is given. To make our presentation clear, some notations are given as follows:

(1) For a binary data sequence where the occurrence ratio of the bits of one and two is the same, based on the required dimming level, we first divide it into a certain number of blocks with binary data length K, denoted as $\mathbf{b} = [b_K, \cdots, b_2, b_1]$, which is taken as the input of the "dimming encoding" part.

(2) The dimming code is denoted as $\mathbf{g} = [\mathbf{g}_K, \cdots, \mathbf{g}_2, \mathbf{g}_1]$, where $\mathbf{g}_i = [g_{iK}, \cdots, g_{i2}, g_{i1}]$ with $i \in \{1, 2, \cdots, K\}$, which corresponds to the specific dimming level. The normalized code weight of dimming code $\mathbf{g}$ is:

$$W = \frac{\sum_{i=1}^{K} \sum_{j=1}^{K} g_{ij}}{K^2}. \qquad (3)$$

(3) We perform the bitwise AND operation on block code $\mathbf{b}$ and each subsection of dimming code $\mathbf{g}_i$ to generate new codeword $\mathbf{c}_i = [c_{iK}, \cdots, c_{i2}, c_{i1}]$. Then, we combine each $\mathbf{c}_i$ to make the transmitted data $\mathbf{c} = [\mathbf{c}_K, \cdots, \mathbf{c}_2, \mathbf{c}_1]$, which satisfies the dimming requirement. For any random

input data sequence, the ratio of bits of zeroes and ones is the same. Namely, these block codes are transmitted with equal probability. Therefore, the dimming level can be expressed as:

$$\gamma = \begin{cases} \frac{W}{2}, & 0 < \gamma \leq 0.5, \\ 1 - \frac{W}{2}, & 0.5 < \gamma < 1. \end{cases} \quad (4)$$

Therefore, the key point of the problem is the construction of the dimming code. To better distinguish the generation of codewords, we provide the following principles to construct the dimming code **g**.

Principle 1. *The construction of the dimming code.*

(1) The dimming code corresponding to the minimum dimming level should guarantee that $g_{ij} = 1$ when $i = j$ and $g_{ij} = 0$ when $i \neq j$.
(2) As mentioned above, the dimming factor can be changed proportionally with the code weight of the dimming code. To increase the dimming level, we can successively increase the number of "ON" levels of each $\mathbf{g}_i$. Therefore, the dimming resolution is $\tilde{\gamma} = \frac{1}{2K^2}$.
(3) The dimming code with $\gamma \in (0.5, 1)$ is the same as the $\gamma \in (0, 0.5)$ part, and accordingly, the generated data are $\tilde{\mathbf{c}}_i = [\tilde{c}_{iK}, \cdots, \tilde{c}_{i2}, \tilde{c}_{i1}]$, where $\tilde{c}_{ik} = 1 - c_{ik}$.

Example 1. *To make the encoding process clear, without loss of generality, we make $K = 3$ here, so the dimming code under different dimming levels is as summarized in Table 1. Meanwhile, the construction of the generated codewords under dimming factor $\gamma = \frac{1}{6}$ with $K = 3$ is shown in Table 2.*

From the above analysis, the transmission efficiency can be expressed as:

$$\nu = \frac{1}{K}, \quad (5)$$

where the transmission rate can be fixed under different dimming levels when the bandwidth is given. We conclude the performance comparison under different K where the corresponding dimming level and dimming range are determined, as shown in Table 3. The analysis shows that we can obtain a wider dimming range and more precise dimming levels with the increase of parameter K. However, a large value of K may degrade the performance of transmission efficiency. Thus, we can choose the proper dimming scheme depending on the specific transmission condition.

Consequently, we can summarize the above-mentioned encoding process as Algorithm 1 where $\otimes$ denotes the operation of bitwise AND.

Algorithm 1 Dimming Encoding

Require:
  dimming factor γ
  input data bit **b** and dimming code **g**
Ensure:
  if $0 < \gamma \leq 0.5$
    $\mathbf{c} = [\mathbf{c}_K, \cdots, \mathbf{c}_2, \mathbf{c}_1]$, where $\mathbf{c}_k = \mathbf{b} \otimes \mathbf{g}_k$
  else $0.5 < \gamma < 1$
    $\mathbf{c} = [\tilde{\mathbf{c}}_K, \cdots, \tilde{\mathbf{c}}_2, \tilde{\mathbf{c}}_1]$, where $\tilde{\mathbf{c}}_k = 1 - \mathbf{c}_k, \mathbf{c}_k = \mathbf{b} \otimes \mathbf{g}_k$

Table 1. The dimming code under different dimming levels with $K = 3$.

g_3	g_2	g_1	γ
100	010	001	$\frac{1}{6}$
110	010	001	$\frac{2}{9}$
111	010	001	$\frac{5}{18}$
111	110	001	$\frac{1}{3}$
111	111	001	$\frac{7}{18}$
111	111	101	$\frac{4}{9}$
111	111	111	$\frac{1}{2}$

Table 2. The construction of generated codewords under dimming factor $\gamma = \frac{1}{6}$ with $K = 3$.

i	b	g_3	g_2	g_1	c
1	000	100	010	001	000000000
2	001	100	010	001	000000001
3	010	100	010	001	000010000
4	100	100	010	001	100000000
5	011	100	010	001	000010001
6	101	100	010	001	100000001
7	110	100	010	001	100010000
8	111	100	010	001	100010001

Table 3. The performance comparison under different K.

K	γ_{min}	γ_{max}	$\tilde{\gamma}$	ν
2	$\frac{1}{4}$	$\frac{3}{4}$	$\frac{1}{8}$	$\frac{1}{2}$
3	$\frac{1}{6}$	$\frac{5}{6}$	$\frac{1}{18}$	$\frac{1}{3}$
4	$\frac{1}{8}$	$\frac{7}{8}$	$\frac{1}{32}$	$\frac{1}{4}$
K	$\frac{1}{2K}$	$\frac{2K-1}{2K}$	$\frac{1}{2K^2}$	$\frac{1}{K}$

3.2. Dimming Decoder

In this subsection, we will demonstrate the process of decoding the structure. To make our presentation clear, some notations are given as follows:

(1) After maximum likelihood (ML) detection, the received binary data can be regarded as the estimation of transmitted data, which is denoted as $\hat{c} = [\hat{c}_K, \cdots, \hat{c}_2, \hat{c}_1]$, where $\hat{c}_i = [\hat{c}_{iK}, \cdots, \hat{c}_{i2}, \hat{c}_{i1}]$ with $i \in \{1, 2, \cdots, K\}$. Then, the binary data sequence $\hat{c}$ is taken as the input of the "dimming decoding" part.

(2) We perform the bitwise OR operation on each subsection of received binary codeword $\hat{c}_i$ to estimate the original input data string $\hat{b}$.

Consequently, we can summarize the above-mentioned encoding process as Algorithm 2 where $\oplus$ denotes the operation of bitwise OR.

Algorithm 2 Dimming Decoding

Require:
received binary data $\hat{c}$
Ensure:
  if $0 < \gamma \leq 0.5$
    $\hat{b} = [\hat{c}_K \oplus \cdots \oplus \hat{c}_2 \oplus \hat{c}_1]$
  else $0.5 < \gamma < 1$
    $\hat{b} = [\overline{\hat{c}_K} \oplus \cdots \oplus \overline{\hat{c}_2} \oplus \overline{\hat{c}_1}]$, where $\overline{\hat{c}_k} = 1 - \hat{c}_k$

By means of the above-mentioned encoding/decoding algorithms, we can simultaneously achieve dimming control and data transmission with a fixed transmission rate.

4. Simulation Results

To evaluate the performance of the proposed encoding and decoding scheme, simulations have been carried out. The indoor environment was designed based on [1] with a room size of 5 m × 5 m × 3 m. The LED was installed at a height of 2.5 m from the floor, and the PD was put on the desk at 0.85 m under the LED. The semi-angle at half-power of the LED chip was 60 deg, and the field of view was 60 deg. The effective detected area of the receiver PD was 1.0 cm^2. The gains of the optical filter and the refractive index of an optical concentrator were set as 1.0 and 1.5, respectively.

Next, to compare the error performance of the various dimming schemes, we fixed the the peak light intensity of transmitted signal P for a fair comparison according to the literature [11,14]. The signal-to-noise ratio (SNR) can be defined as $\text{SNR} = 10\log_{10}[\frac{1}{R_c}\frac{HP}{\sigma^2}]$, where the code rate R_c can be regarded as transmission efficiency here, and σ^2 is the variance of the AWGN. The bit error rate (BER) was calculated by the Monte Carlo method, and the length of the transmitted data was set to be 10^8 in the simulations. Simulations of BER have been carried out for the proposed dimming scheme with the different dimming levels where the SNR gain (in dB) was utilized, and three representative dimming levels (i.e., 0.25, 0.375, and 0.5) with different parameter K values were considered to examine the performance of the proposed scheme.

As shown in Figure 3, the red lines and blue lines represent the proposed scheme with parameter values $K = 2$ and $K = 4$ respectively. As the parameter K increased, the code rate R_c decreased accordingly. We can see that the BER performance with $K = 2$ outperformed $K = 4$ under the same dimming levels. In terms of the same parameter $K = 2$, the error performance with dimming level $\gamma = 0.25$ was better than dimming levels $\gamma = 0.375$ and $\gamma = 0.5$, 1.7 dB and 2.7 dB SNR gains at BER $= 10^{-4}$, respectively.

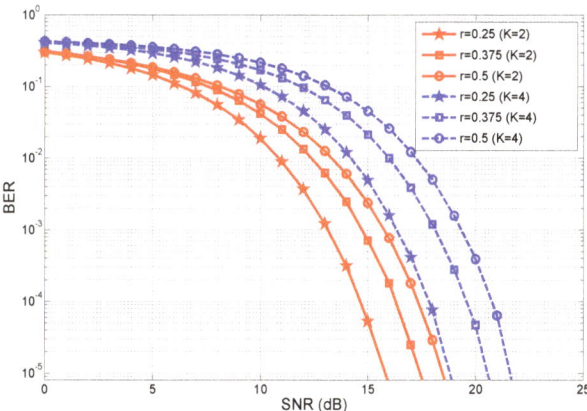

Figure 3. Error performance of the proposed scheme under different dimming factor.

5. Conclusions

In this paper, we have proposed a block coding-based scheme to achieve dimming control, as well as data transmission with constant transmission efficiency, where a large dimming range can be achieved. Meanwhile, the encoding/decoding algorithm was provided accordingly. Via the bitwise AND operation on the block code and each subsection of dimming code to construct the proper codewords satisfying the dimming requirement, the proposed scheme is simple to implement and can maintain constant transmission efficiency under different dimming levels. Simulation results show that the proposed scheme can achieve reliable data transmission via the designed encoding/decoding structure. Therefore, it proves an attractive alternative dimming scheme.

Author Contributions: All authors contributed equally to writing the paper. Formal analysis, Y.Z.; funding acquisition, J.Z.; writing, original draft, Y.Z.; writing, review and editing, J.Z.

Funding: This work is supported in part by Grant No. 161100210200 from the Major Scientific and Technological of Henan Province, China.

Acknowledgments: The authors wish to thank the anonymous reviewers for their valuable suggestions.

Conflicts of Interest: The authors declare no conflict of interest.

References

1. Komine, T.; Nakagawa, M. Fundamental analysis for visible light communication system using LED lights. *IEEE Trans. Consum. Electron.* **2004**, *50*, 100–107. [CrossRef]
2. Vucic, J.; Kottke, C.; Nerreter, S.; Langer, K.D.; Walewski, J.W. 513 mbit/s visible light communications link based on dmt-modulation of a white LED. *IEEE J. Lightw. Technol.* **2010**, *28*, 3512–3518. [CrossRef]
3. Karunatilaka, D.; Zafar, F.; Kalavally, V.; Parthiban, R. Led based indoor visible light communications: State of the art. *IEEE Commun. Surv. Tutor.* **2015**, *17*, 1649–1678. [CrossRef]
4. Jovicic, A.; Li, J.; Richardson, T. Visible light communication: Opportunities, challenges and the path to market. *IEEE Commun. Mag.* **2013**, *51*, 26–32. [CrossRef]
5. Randel, S.; Breyer, F.; Lee, S.C.J.; Walewski, J.W. Advanced modulation schemes for short-range optical communications. *IEEE J. Sel. Top. Quant.* **2010**, *16*, 1280–1289. [CrossRef]
6. Zafar, F.; Karunatilaka, D.; Parthiban, R. Dimming schemes for visible light communication: The state of research. *IEEE Wirel. Commun.* **2015**, *22*, 29–35. [CrossRef]
7. Rajagopal, S.; Roberts, R.D.; Lim, S.K. IEEE 802.15.7 visible light communication: Modulation schemes and dimming support. *IEEE Commun. Mag.* **2012**, *50*, 72–82. [CrossRef]
8. Lee, K.; Park, H. Modulations for visible light communications with dimming control. *IEEE Photonics Technol. Lett.* **2011**, *23*, 1136–1138. [CrossRef]
9. Sang, H.L.; Jung, S.Y.; Kwon, J.K. Modulation and coding for dimmable visible light communication. *IEEE Commun. Mag.* **2015**, *53*, 136–143. [CrossRef]
10. Kim, J.; Park, H. A coding scheme for visible light communication with wide dimming range. *IEEE Photonics Technol. Lett.* **2014**, *26*, 465–468. [CrossRef]
11. Kim, S.; Jung, S.Y. Novel FEC coding scheme for dimmable visible light communication based on the modified reedmuller codes. *IEEE Photonics Technol. Lett.* **2011**, *23*, 1514–1516. [CrossRef]
12. Sang, H.L.; Ahn, K.I.; Kwon, J.K. Multilevel transmission in dimmable visible light communication systems. *IEEE J. Lightw. Technol.* **2013**, *31*, 3267–3276. [CrossRef]
13. Lee, K.; Park, H.; Barry, J.R. Indoor channel characteristics for visible light communications. *IEEE Commun. Lett.* **2011**, *15*, 217–219. . [CrossRef]
14. Zuo, Y.; Zhang, J.; Zhang, Y.Y.; Chen, R.H. Weight threshold check coding for dimmable indoor visible light communication systems. *IEEE Photonics J.* **2018**, *10*, 1–11. [CrossRef]

© 2019 by the authors. Licensee MDPI, Basel, Switzerland. This article is an open access article distributed under the terms and conditions of the Creative Commons Attribution (CC BY) license (http://creativecommons.org/licenses/by/4.0/).

Article

Low-Complexity Hybrid Optical OFDM with High Spectrum Efficiency for Dimming Compatible VLC System

Simeng Feng [1], Hailiang Feng [2], Ying Zhou [3] and Baolong Li [3,*]

1. Next Generation Wireless, University of Southampton, Southampton SO17 1BJ, UK
2. Academy of Broadcasting Science, National Radio and Television Administration, Beijing 100866, China
3. Jiangsu Provincial Engineering Laboratory of Pattern Recognition and Computational Intelligence, Jiangnan University, Wuxi 214122, China
* Correspondence: lblong@jiangnan.edu.cn; Tel.: +86-134-5181-4793

Received: 31 July 2019; Accepted: 2 September 2019; Published: 4 September 2019

Abstract: In visible light communications (VLC), dimming control constitutes an indispensable technique to comply with various illumination necessities and with different energy consumption constraints. Therefore, a novel dimming compatible hybrid optical orthogonal frequency division multiplexing (DCHO-OFDM) is conceived in this paper to fulfil the requirements from communications and illuminations. Explicitly, the signal branch of the unclipped asymmetrically clipped O-OFDM (ACO-OFDM) and the down/upper-clipped pulse-amplitude-modulated discrete multitone (PAM-DMT) are adaptively amalgamated in order to increase the spectrum efficiency. For the sake of precisely achieving dimming control, the chromaticity-shift-free and industry-preferred pulse width modulation (PWM) is further invoked to the hybrid signal, assisted by a time-varying biasing scheme to mitigate the non-linear distortion. As the different signal components in DCHO-OFDM are beneficially combined in an interference-orthogonal approach, the transmitted symbols are able to be readily detected upon relying on a standard OFDM receiver, as that of ACO-OFDM. Our simulations demonstrate that a high spectrum efficiency of the conceived DCHO-OFDM scheme can be achieved with less fluctuation in a wide dimming range.

Keywords: visible light communications (VLC); optical orthogonal frequency division multiplexing (O-OFDM); dimming control; pulse width modulation (PWM)

1. Introduction

Driven by the rapid development of light emitting diodes (LED), visible light communication (VLC) constitutes a promising part of next-generation wireless communications [1–3]. Under the trend of internet of everything (IoE), the exhaustion of radio frequency (RF) band to resolve the unprecedented data escalation dilemma becomes a bottleneck of current communications [4,5]. As a benefit, the environmentally friendly technology of VLC possesses plenty of unlicensed spectral resource at the high frequency band [6]. This part of electromagnetic-interference-free spectrum can be exploited, for the sake of supporting high-speed and high-security communications in the near future [7]. Upon relying on modulating the intensity of lights above the flicker-fusion frequency, the dual functionality of communications and illuminations can be simultaneously achieved in VLC, providing an efficient solution especially for short-range communications [8].

In order to achieve high-speed transmission, optical orthogonal frequency division multiplexing (O-OFDM) modulation schemes have been extensively explored for VLC systems [9]. Since a VLC system is based on intensity modulation and direct detection (IM/DD), the transmitted signal via VLC link should be both real-valued and positive-valued [10]. To meet these constraints, there are two classic O-OFDM schemes in VLC, namely, asymmetrically clipped O-OFDM (ACO-OFDM) and DC-biased O-OFDM (DCO-OFDM). DCO-OFDM relies on a DC bias to guarantee the positivity, which enjoys the advantage of implementation simplicity. However, it suffers from the power inefficiency since the DC bias which does not convey any information leads to a waste of power. By contrast, with the aid of the properties of Fourier transform, ACO-OFDM directly clips the negative parts of the signal to generate a non-negative signal, substantially increasing the power efficiency, but ACO-OFDM leverages only half of the subcarriers for transmission hence, resulting in spectrum inefficiency [11]. Therefore, advanced transmission schemes have been further investigated, such as hybrid ACO-OFDM (HACO-OFDM) [12] and layered ACO-OFDM (LACO-OFDM) [13]. In HACO-OFDM regimes, for the sake of promoting the spectrum efficiency, the ACO-OFDM signal is further combined with pulse-amplitude-modulated discrete multitone (PAM-DMT), where the latter is employed to modulate the imaginary part of even-indexed subcarriers. Superior to the conventional OFDM schemes, HACO-OFDM can notably improve the spectrum efficiency while, at the same time, maintaining high power efficiency. However, on the receiver side of HACO-OFDM, additional operations including the re-generation of time-domain ACO-OFDM samples, the zero-clipping, the re-production of clipping distortion and the noise subtraction are required, leading to an increased complexity [14]. The perspective of LACO-OFDM improves the traditional ACO-OFDM upon relying on actively invoking the unexploited subcarriers for transmission in a layer-based approach. Compared to HACO-OFDM, the spectrum efficiency can be further enhanced. Unfortunately, the improvement of spectrum efficiency in LACO-OFDM is accompanied by the complexity increment due to the multi-layer transmitter and the successive interference cancellation (SIC) receiver [15].

It is worth mentioning that conventional O-OFDM schemes mainly concentrate on promoting data rate, while ignoring the various illumination requirements. As an emerging member of green communication technology, VLC is intrinsically distinct from the conventional RF system due to its dual functionality: communications and ambient lighting, and not just the former [16]. Therefore, the dimming control is envisioned to be a critical technique in VLC, for the sake of handling the lighting and energy consumption constraints [17]. There are mainly two kinds of dimming approaches extensively used in VLC, namely, analogous dimming and digital dimming. A popular analogous dimming approach, known as the conventional DCO-OFDM, achieves dimming control upon adjusting the DC bias, which is easily implemented. Unfortunately, the performance of DCO-OFDM may fluctuate with different dimming levels. Especially when the requested dimming level is considerably low or high, the resulting performance becomes degraded, due to the grave non-linear distortion [18]. The analogous dimming approach is further investigated in terms of advanced O-OFDM scheme, such as the asymmetrical hybrid optical-OFDM (AHO-OFDM) scheme [19]. In AHO-OFDM, either the signal of PAM-DMT or ACO-OFDM is inverted, and then compounded for transmission. By modifying the power allocated to the two components, a wide dimming range can be attained. However, the analogous dimming approach is challenging to precisely control the brightness and may further impinge upon the wavelength of the emitted light, resulting in undesired chromaticity shift [20]. Therefore, another category of the dimming approach, which is referred to as digital dimming relying on the the pulse width modulation (PWM) technique has been extensively applied. A reverse polarity O-OFDM (RPO-OFDM) was proposed to dim the transmitted signal, such as the conventional ACO-OFDM, upon invoking the PWM technique [21]. RPO-OFDM can realize a steady BER performance over a wide range of dimming levels, but has a low spectrum efficiency. To improve the spectrum efficiency, a negative HACO-OFDM (NHACO-OFDM) scheme was conceived in [22] for combining with the high level of PWM signal, which is further amalgamated with HACO-OFDM at the low level. In this regime, the various dimming requests can be satisfied upon altering the proportion of these two branches, while a high spectral efficiency can be realized. Moreover, the authors in [23] proposed a reconstructed layered ACO-OFDM (RLACO-OFDM) scheme based on LACO-OFDM, which is further combined with PWM signal to achieve various brightness levels, without introducing additional complexity. As a benefit, the PWM-biased dimming approach constitutes an implementation-friendly philosophy, where the brightness level can be well-controlled upon varying the parameter of duty cycle [24].

However, the existing digital dimming approaches are vulnerable to the varying dimming level, where a preliminary is required at the receiver side to detect the PWM signal, leading to an increasing complexity. In fact, detecting the PWM signal may be intractable especially in the scenario of multiple LEDs, since the received multi-path signals may have different dimming levels [23]. Furthermore, as mentioned above, both the HACO-OFDM and LACO-OFDM require a successive demodulation operation, suffering from an increased receiver complexity. Against this background, *a novel dimming compatible hybrid optical OFDM (DCHO-OFDM) is conceived in this paper incorporating the PWM technique to realize high-spectral-efficiency transmission under various dimming levels, where the transmitted signal can be readily recovered upon relying on a standard receiver.* To be specific,

- The conceived DCHO-OFDM signal is beneficially composed of the unclipped ACO-OFDM signal and the down/upper-clipped unipolar PAM-DMT signal, where the time-varying bias is further facilitated to diminish the undesired non-linear distortion. In this way, the proposed DCHO-OFDM achieves a higher spectrum efficiency than the conventional schemes over a broad dimming range.
- Thanks to the elaborately combined signal, the symbol transmitted by the proposed DCHO-OFDM is capable of being recovered without detecting the PWM signal, which is superior to the existing digital dimming approaches. Additionally, the BER performance of DCHO-OFDM is invulnerable to the various duty cycles of PWM, achieving a steady system performance in a wide dimming range compared to DCO-OFDM.
- Moreover, the information-carried subcarriers in the proposed DCHO-OFDM scheme can avoid the corruption by the generated interference. As a benefit, the symbols transmitted by the proposed DCHO-OFDM can be readily detected upon relying on a standard OFDM receiver, which significantly reduces the complexity at receiver compared to the existing advanced transmission schemes, such as HACO-OFDM and LACO-OFDM.

The superiorities of the proposed DCHO-OFDM in terms of spectrum efficiency over its counterparts have been verified by our simulation results.

The rest of this paper is organized as follows. Section 2 provides a review of the HACO-OFDM scheme in VLC. The detail of the proposed DCHO-OFDM regime is presented in Section 3. Our numerical results are discussed in Section 4, whilst our conclusions are drawn in Section 5.

2. Brief Review of HACO-OFDM

To enhance the spectrum efficiency, HACO-OFDM scheme amalgamates the signal of ACO-OFDM with PAM-DMT for achieving an ambitious transmission rate [12]. In order to guarantee real-valued outputs, Hermitian symmetry is invoked in each branch. As demonstrated in Figure 1a, the mapped M-ary quadrature amplitude modulation (M-QAM) symbols A_i in ACO-OFDM branch occupy the odd-indexed subcarriers, which is expressed as:

$$\mathbf{X} = [0, A_1, 0, A_2, \cdots, A_{N/4}, 0, A^*_{N/4}, \cdots, A^*_2, 0, A^*_1], \quad (1)$$

where the total number of subcarriers is denoted by N and A^*_i is invoked to present the Hermitian symmetry of symbol A_i. After performing the N-point IFFT of $\mathbf{X}$, a bipolar time-domain signal $x_{ACO,n}$ is obtained. In order to acquire the positive signal appropriating for VLC propagation, the bipolar signal of $x_{ACO,n}$ has to be directly clipped at zero as:

$$\lfloor x_{ACO,n} \rfloor_c = \begin{cases} 0, & x_{ACO,n} < 0 \\ x_{ACO,n}, & x_{ACO,n} \geq 0 \end{cases}, \quad (2)$$

where the resultant clipping distortion is imposed on the even-indexed subcarriers, leaving the odd-indexed subcarriers carrying ACO-OFDM symbols uncontaminated [25]. For the PAM-DMT link, as depicted in Figure 1a, the mapped L-ary pulse amplitude modulation (L-PAM) symbols B_i are assigned to the the even-indexed subcarriers, with only the imaginary component being modulated. It can be expressed as:

$$\mathbf{Y} = \left[0, 0, jB_1, 0, jB_2, \cdots, jB_{N/4-1}, 0, 0, 0, jB^*_{N/4-1}, 0, \cdots, jB^*_2, 0, jB^*_1, 0\right], \tag{3}$$

where we have $j = \sqrt{-1}$. A bipolar signal of $y_{PAM,n}$ can be then obtained, upon performing the N-point IFFT operation of $\mathbf{Y}$. For the sake of producing the non-negative result, $y_{PAM,n}$ is clipped at zero-level, which is given by:

$$\lfloor y_{PAM,n} \rfloor_c = \begin{cases} 0, & y_{PAM,n} < 0 \\ y_{PAM,n}, & y_{PAM,n} \geq 0 \end{cases}. \tag{4}$$

Consequently, in HACO-OFDM scheme, the clipped signal $\lfloor x_{ACO,n} \rfloor_c$ and $\lfloor y_{PAM,n} \rfloor_c$ is capable of being simultaneously transmitted, achieving a beneficially high spectrum efficiency. As the detrimental clipping noise generated by ACO-OFDM contaminates the subcarriers carrying PAM-DMT signal, the receiver of HACO-OFDM has to recover the ACO-OFDM symbols first and then retrieve the clipping distortion imposed on the even-indexed subcarriers, so that the detection of PAM-DMT signal can be successively implemented.

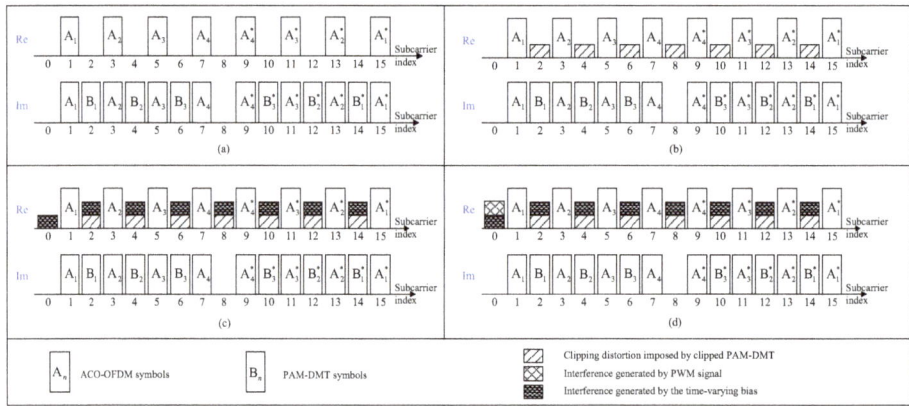

Figure 1. Frequency-domain demonstration including both the real component and the imaginary component of the proposed DCHO-OFDM signal, with $N = 16$ subcarriers. (**a**) presents the symbol allocation for ACO-OFDM and for PAM-DMT branch, respectively; (**b**) demonstrates the hybrid unclipped ACO-OFDM and the down/upper-clipped PAM-DMT signals; (**c**) indicates the combined signal after performing time-varying biasing scheme; (**d**) depicts the hybrid signal further incorporating the PWM signal.

For achieving dimming control, directly combining the HACO-OFDM signal with PWM is deemed to be inefficient. Due to the non-negativity of the clipped unipolar signal $\lfloor x_{ACO,n} \rfloor_c$ and $\lfloor y_{PAM,n} \rfloor_c$, the hybrid signal may exceed the dynamic range of LEDs with a large probability [26]. To overcome, a negative-valued NHACO-OFDM signal is invoked to superpose to the high brightness level of PWM signal to avoid non-linear distortion, while the HACO-OFDM with positive-polarity is invoked for the low-brightness scenario [22]. In this regime, the PWM signal has to be detected first by a receiver in order to recover the dissimilar clipping distortion generated by NHACO-OFDM and HACO-OFDM, respectively. Afterwards, the transmitted information can be then extracted by successively demodulating the ACO-OFDM and PAM-DMT branch.

3. Proposed Dimming Compatible Hybrid O-OFDM

Although the spectrum efficiency is improved in HACO-OFDM, the sequential detection operations lead to an increased receiver complexity and process delay, compared to that of the conventional ACO/DCO-OFDM. When dimming control is invoked, additional operation including the detection of PWM signal is necessary before demodulation, which further increases the implementation complexity. To overcome this, in this paper, a novel dimming compatible hybrid O-OFDM (DCHO-OFDM) scheme is conceived with a low-complexity receiver for achieving a high-speed VLC transmission. Moreover, the proposed scheme is insensitive to the different dimming levels, which implies the preliminary of detecting PWM signal can be neglected.

3.1. Pulse Width Modulation

Due to the non-linear transfer characteristic of LED, the transmitted VLC signal has to be confined in the limited linear range. We employ the notation I_l and I_h to represent the lower and upper bound of the limited linear range, respectively. Then, the dimming level η can be defined as:

$$\eta = \frac{I_{ave} - I_l}{I_h - I_l}, \qquad (5)$$

where we have I_{ave} to present the average amplitude of signal. Upon observing Equation (5), different dimming levels can be attained by altering the value of I_{ave}. Although directly adding a bias can change the value of I_{ave}, it leads to an undesired wavelength shift of the emitted lights. Therefore, the PWM signal is invoked in our proposed DCHO-OFDM scheme, which can be expressed as:

$$d_{PWM}(t) = \begin{cases} I_h, & 0 < t \leq T_{ON} \text{ (ON state)} \\ I_l, & T_{ON} < t \leq T \text{ (OFF state)} \end{cases} \qquad (6)$$

where T represents the period of PWM signal and T_{ON} denotes the time duration when the signal is in the ON state. The dimming level can be changed upon adjusting the duty cycle of PWM, which is referred to as $\rho = T_{ON}/T$.

3.2. DCHO-OFDM

In order to implement dimming control, the PWM signal needs to be designated for incorporating the O-OFDM transmission. Specifically, the period of the PWM is set to be as many as the K times of a single O-OFDM symbol duration T_s, which can be represented as $T = KT_s$, where K is an integer. Accordingly, the duration of ON state of PWM is set to $T_{ON} = K'T_s$, where K' denotes the number of O-OFDM symbols transmitted at ON state. It can be observed that the PWM signal remains constant in a single O-OFDM symbol duration T_s, which implies that the introduced interference is imposed only on the 0-th subcarrier, as depicted in Figure 1d.

Since the transmitted signal in VLC has to be constrained by the limited linear transmission range of LEDs, the O-OFDM components needs to be carefully designed, in order to incorporate PWM for the purpose of dimming control. Taking the characteristic of the PWM signal into consideration, a signal with negative polarity is desired during the ON state, while a positive-valued signal is required for the OFF state. Therefore, instead of combining the unipolar signal of ACO-OFDM and of PAM-DMT as seen in the conventional HACO-OFDM scheme, we invoke the unclipped double-side ACO-OFDM signal in our proposed DCHO-OFDM to amalgamated with either the upper-clipped PAM-DMT signal $\lceil y_{PAM,n} \rceil_c$ or the down-clipped PAM-DMT signal $\lfloor y_{PAM,n} \rfloor_c$ as:

$$\begin{cases} s_n^{on} = x_{ACO,n} + \lceil y_{PAM,n} \rceil_c, & \text{ON state} \\ s_n^{off} = x_{ACO,n} + \lfloor y_{PAM,n} \rfloor_c, & \text{OFF state} \end{cases}, \qquad (7)$$

depending on the state of PWM signal. Note that the upper-clipped PAM-DMT signal is given by:

$$\lceil y_{PAM,n} \rceil_c = \begin{cases} 0, & y_{PAM,n} \geq 0 \\ y_{PAM,n}, & y_{PAM,n} < 0 \end{cases}. \qquad (8)$$

Similar to the HACO-OFDM scheme, the ACO-OFDM symbols in our DCHO-OFDM are allocated to the odd-indexed subcarriers, where the PAM-DMT symbols are carried by the imaginary proportion of the even-indexed subcarriers, as shown in (1) and (3). On the contrary, the ACO-OFDM branch in the proposed DCHO-OFDM scheme is designed to remain unclipped, yielding even-indexed subcarriers that are free of clipping noise contamination. According to [27], the clipping distortion generated by the down-clipped PAM-DMT is imposed only on the real-part of the even-indexed subcarriers. Furthermore, upon the analysis provided in Appendix A, the clipping distortion caused by the upper-clipped PAM-DMT is also encountered only by the real-portion of the even-indexed subcarriers. Therefore, it implies that the overall generated clipping distortion in DCHO-OFDM of either the combined signal s_n^{on} or s_n^{off} is orthogonal to the information-carrying subcarriers, which is demonstrated in Figure 1b. As a benefit, the bipolar signal of s_n^{on} and s_n^{off} can be obtained in an interference-orthogonal way. Furthermore, in HACO-OFDM scheme, the clipped ACO-OFDM and PAM-DMT branch are superposed directly, which leads to a higher amplitude since the both components are non-negative. In contrast, by superposing the bipolar ACO-OFDM and the unipolar PAM-DMT signal, a lower peak-to-average-power-ratio (PAPR) can be expected, compared to that of the conventional HACO-OFDM.

The hybrid signal in Equation (7) has double-sided polarity, as seen in Figure 2a,b. In order to incorporate the PWM signal, it has to be converted to the unipolar signal. To tackle this issue, we introduce a time-varying biasing scheme based on the Proposition 1, which converts the the bipolar signal of s_n^{on} and s_n^{off} to negative-valued and positive-valued signals for the ON and OFF states, respectively. Moreover, the time-varying bias does not induce any additional interference imposed on the legitimate transmitted symbols. To be specific, the Proposition 1 is exhibited as follows.

Proposition 1. *If a time-domain signal z_n has the characteristic as:*

$$\begin{cases} z_n = z_{\frac{N}{2}+n}, & n = 0, \frac{N}{4} \\ z_n = z_{\frac{N}{2}-n} = z_{\frac{N}{2}+n} = z_{N-n}, & n = 1, \cdots, \frac{N}{4} - 1 \end{cases}, \qquad (9)$$

the interference generated by z_n imposes only on the real-part of the even-indexed subcarriers.

Proof. Please see Appendix B. □

For the hybrid signal s_n^{on}, it is expected to be negative for further incorporating the PWM signal during ON state. In the light of Proposition 1, a bias of $-\max\left\{s_n^{on}, s_{N/2+n}^{on}\right\}$ can be added to sample s_n^{on} and $s_{N/2+n}^{on}$, for $n = 0, N/4$, so that to achieve the negative-valued samples. Uniformly, a bias of $-\max\left\{s_n^{on}, s_{N/2-n}^{on}, s_{N/2+n}^{on}, s_{N-n}^{on}\right\}$ is applied for sample $s_n^{on}, s_{N/2-n}^{on}, s_{N/2+n}^{on}$, and s_{N-n}^{on} to acquire the non-positive values, when we have $n = 1, 2, \cdots, N/4 - 1$. Therefore, the introduced time-varying bias z_n^{on} for signal s_n^{on} is given by:

$$\begin{cases} z_n^{on} = z_{\frac{N}{2}+n}^{on} = -\max\left\{s_n^{on}, s_{\frac{N}{2}+n}^{on}\right\}, & n = 0, \frac{N}{4} \\ z_n^{on} = z_{\frac{N}{2}-n}^{on} = z_{\frac{N}{2}+n}^{on} = z_{N-n}^{on} \\ \quad = -\max\left\{s_n^{on}, s_{\frac{N}{2}-n}^{on}, s_{\frac{N}{2}+n}^{on}, s_{N-n}^{on}\right\}, & n = 1, \cdots, \frac{N}{4} - 1 \end{cases}. \quad (10)$$

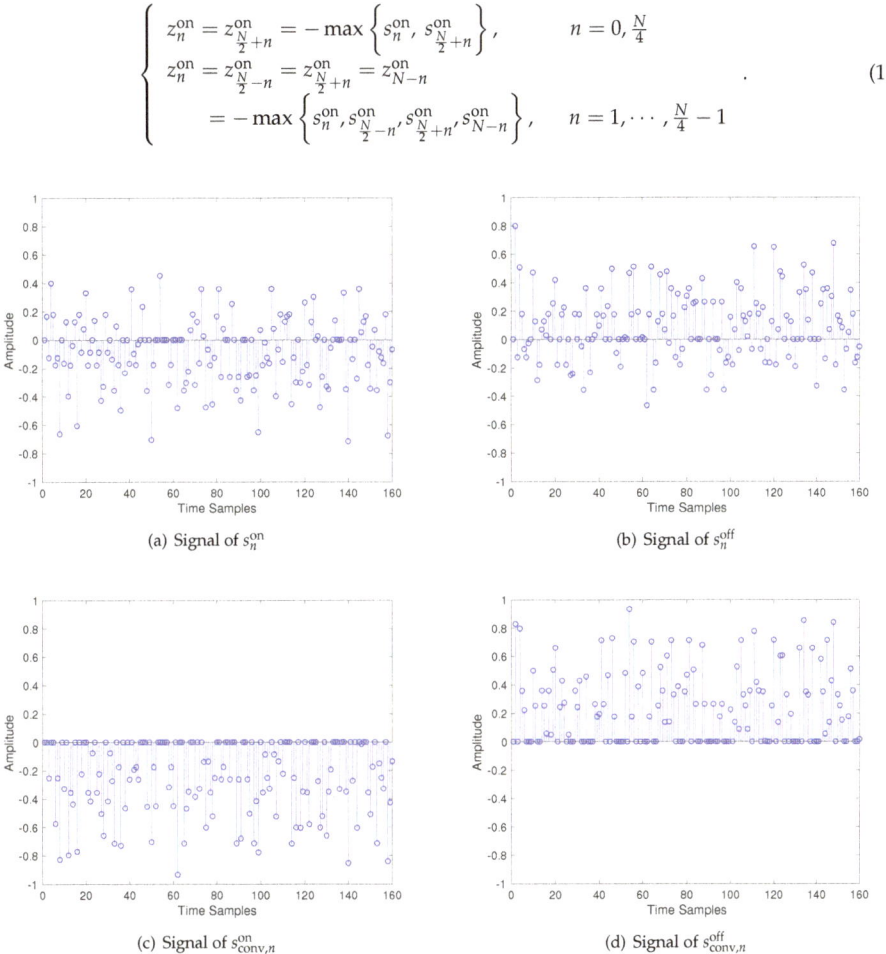

Figure 2. A demonstration of the superposed signal in DCHO-OFDM. (**a**,**b**) depict the bipolar signal s_n^{on} and s_n^{off}, where the ACO-OFDM signal is directly combined with upper/down-clipped PAM-DMT; (**c**,**d**) present the unipolar signal of $s_{conv,n}^{on}$ and $s_{conv,n}^{off}$, aided by the introduced time-varying biasing scheme.

After performing the time-varying biasing, the bipolar signal s_n^{on} is able to be converted to the unipolar signal as:

$$s_{conv,n}^{on} = x_{ACO,n} + \lceil y_{PAM,n} \rceil_c + z_n^{on}, \quad n = 0, 1, \cdots, N-1, \quad (11)$$

where all the N samples experience non-positive values, which is demonstrated in Figure 2c. On the contrary, a positive-valued signal is desired during OFF state transmission. Therefore, for the bipolar signal s_n^{off}, the time-varying bias z_n^{off} is introduced as:

$$\begin{cases} z_n^{\text{off}} = z_{\frac{N}{2}+n}^{\text{off}} = -\min\left\{s_n^{\text{off}}, s_{\frac{N}{2}+n}^{\text{off}}\right\}, & n = 0, \frac{N}{4} \\ z_n^{\text{off}} = z_{\frac{N}{2}-n}^{\text{off}} = z_{\frac{N}{2}+n}^{\text{off}} = z_{N-n}^{\text{off}} \\ \quad = -\min\left\{s_n^{\text{off}}, s_{\frac{N}{2}-n}^{\text{off}}, s_{\frac{N}{2}+n}^{\text{off}}, s_{N-n}^{\text{off}}\right\}, & n = 1, \cdots, \frac{N}{4} - 1 \end{cases} \quad (12)$$

Hence, the converted signal is given by:

$$s_{\text{conv},n}^{\text{off}} = x_{ACO,n} + \lfloor y_{PAM,n} \rfloor_c + z_n^{\text{off}}, \quad n = 0, 1, \cdots N - 1, \quad (13)$$

resulting in a positive-valued signal as shown in Figure 2d, which is appropriated for OFF state transmission.

Assisted by the time-varying biasing scheme, the acquired unipolar signals relying on the Equations (11) and (13) can be readily amalgamated with PWM signal. To this end, the proposed DCHO-OFDM signal can be expressed as:

$$\begin{aligned} v_n &= \begin{cases} s_{\text{conv},n}^{\text{on}} + I_h, & \text{ON state} \\ s_{\text{conv},n}^{\text{off}} + I_l, & \text{OFF state} \end{cases} \\ &= s_{\text{conv},n} + d_{PAM,n}, \quad n = 0, 1, \cdots, N - 1. \end{aligned} \quad (14)$$

Note that the signal of $s_{\text{conv},n}$ consists of the ON state transmitted signal, $s_{\text{conv},n}^{\text{on}}$, and the OFF state transmitted signal, $s_{\text{conv},n}^{\text{off}}$, while $d_{PAM,n}$ is the sample of PWM signal $d_{PAM}(t)$. It is observed from Equation (14) that the PWM dimming control and the OFDM transmission can be well integrated in the conceived DCHO-OFDM. Specifically, Figure 3 depicts the block diagram of the proposed DCHO-OFDM.

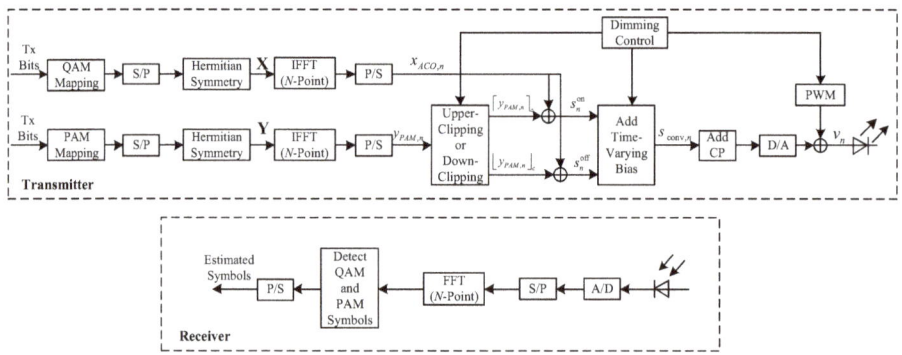

Figure 3. Block diagram of the transmitter and receiver for the proposed DCHO-OFDM.

Furthermore, according to Equation (14), the average amplitude of signal v_n is given by:

$$\begin{aligned} I_{ave} &= \frac{T_{ON}}{T}\left(I_h + \mathbb{E}\left[s_{\text{conv},n}^{\text{on}}\right]\right) + \frac{T - T_{ON}}{T}\left(I_l + \mathbb{E}\left[s_{\text{conv},n}^{\text{off}}\right]\right) \\ &= I_l + E_1 + \rho(I_h - I_l + E_2 - E_1) \end{aligned} \quad (15)$$

where $\mathbb{E}[\bullet]$ denoted the expectation operation. Here, we have E_1 and E_2 to denote the expectation result of signal $s_{\text{conv},n}^{\text{off}}$ and $s_{\text{conv},n}^{\text{on}}$, respectively, which can be readily obtained by numerical simulations. Recalling the Equation (5), the dimming level can then be further expressed as:

$$\eta = \frac{E_1}{I_h - I_l} + \rho\left[1 + \frac{E_2 - E_1}{I_h - I_l}\right]. \quad (16)$$

The Equation (16) shows that the dimming level η has a linear relationship with the duty cycle ρ.

It is essential to discuss the receiver structure for the proposed DCHO-OFDM scheme. The clipping distortion caused by down/upper-clipped PAM-DMT branch in DCHO-OFDM is orthogonal to the information-carrying subcarriers. Moreover, in the light of Proposition 1, the odd-indexed subcarriers carrying ACO-OFDM and the imaginary-part of the even-indexed subcarriers carrying PAM-DMT is capable of remaining uncontaminated during the polarity conversion process relying on the introduced time-varying bias, which is demonstrated in Figure 1c. When the converted signal is amalgamated with PWM for dimming control, the interference caused by PWM falls only on the 0-th subcarrier, which does not introduce additional interference to the transmitted signal. As a benefit, the DCHO-OFDM signal is obtained in an interference-orthogonal approach. Therefore, DCHO-OFDM symbols can be readily recovered upon relying on a standard OFDM receiver, as shown in Figure 3.

3.3. Complexity Analysis

In this section, the complexity of the proposed DCHO-OFDM is discussed. For comparison, the conventional HACO-OFDM is invoked as a benchmarker. At the transmitter side, the main difference between the conventional HACO-OFDM and the proposed scheme lies in the signal component. To elaborate, the zero-clipping operation of the ACO-OFDM branch is required in HACO-OFDM, while the time-varying biasing process is introduced in the proposed DCHO-OFDM. Remarkably, the zero-clipping operation of ACO-OFDM branch is based on computing the maximum value between zero and the signal sample, which is similar to the calculation process of the time-varying biasing scheme in the proposed DCHO-OFDM. Therefore, the proposed DCHO-OFDM is capable of achieving the similar complexity as the conventional HACO-OFDM at the transmitter.

At the receiver side, due to the undesired clipping distortion generated by the ACO-OFDM branch, an additional process including the clipping noise re-generation and cancellation is required in HACO-OFDM. Therefore, the associated computational complexity mainly relies on the sum of two real-valued N-point FFT operations and of one complex-valued N-point IFFT operation [14]. On the contrary, a benefit to the orthogonal interference, the detection of DCHO-OFDM, can be achieved upon relying on a standard OFDM receiver seen in Figure 3, as that of ACO/DCO-OFDM. Hence, the computational complexity of DCHO-OFDM is dominated by a single real-valued N-point FFT operation, which is significantly reduced compared to the HACO-OFDM counterpart [28]. Moreover, for the HACO-OFDM-based dimming control scheme [22], an additional operation is required at the receiver side to distinguish if either HACO-OFDM or NHACO-OFDM is transmitted, since they introduce different clipping distortions. Instead, the proposed DCHO-OFDM signal is invulnerable to the varying dimming level and can be directly recovered without detecting the PWM signal, which futher reduces the complexity compared to that of the HACO-OFDM scheme.

To this end, superior to the conventional HACO-OFDM scheme, the complexity and the process delay of the proposed DCHO-OFDM scheme at the receiver is significantly reduced, without increasing the complexity at the transmitter side.

4. Numerical Results

This section provides our simulations for quantifying the proposed DCHO-OFDM, where we employ $N = 256$ for the O-OFDM VLC system. The lower bound and the upper bound of LED are referred to as $I_l = 0$ and $I_h = 1$, respectively. In order to evaluate the non-linear distortion, a pair of scaling factors β_{ACO} and β_{PAM} are defined as $\beta_{ACO} = (I_h - I_l)/\sigma_{ACO}$ and $\beta_{PAM} = (I_h - I_l)/\sigma_{PAM}$, where σ_{ACO} and σ_{PAM} denote the standard variance of the unclipped ACO-OFDM and PAM-DMT, respectively.

To begin with, we investigate the PAPR performance of the proposed DCHO-OFDM scheme, compared to that of the HACO-OFDM, where the associated complimentary cumulative distribution function (CCDF) can be seen in Figure 4 [26]. It is observed that the proposed DCHO-OFDM achieves a much lower PAPR than that of the conventional HACO-OFDM. This observation is as expected for the reason that the unclipped ACO-OFDM is invoked in DCHO-OFDM, which is capable of avoiding relatively high amplitudes. Note that a lower PAPR results in a better ability to resist the non-linearity [26]. Therefore, unlike the HACO-OFDM, the proposed OFDM scheme is capable of alleviating the undesired non-linear distortion in the presence of the non-linear transfer characteristic of LED. In this case, a better BER performance can be achieved by the proposed DCHO-OFDM, leading to an improved achievable spectrum efficiency compared to HACO-OFDM, as shown in the following figures.

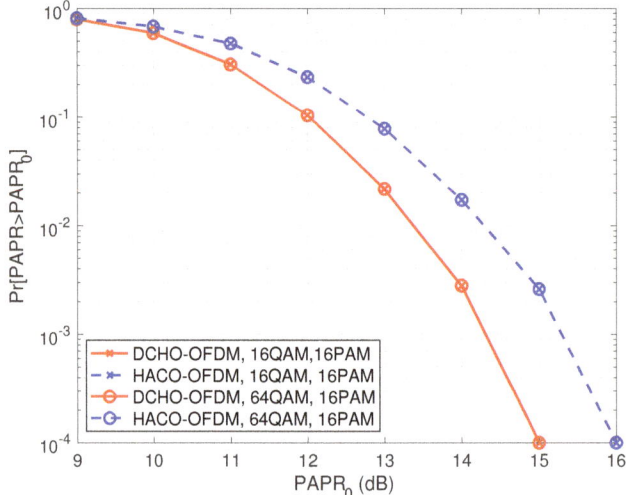

Figure 4. CCDF curves of PAPR for the proposed DCHO-OFDM scheme, compared to that of the HACO-OFDM scheme.

Figure 5 quantifies the dimming level for the various values of duty cycle ρ. Upon observing, the dimming level is linearly changed upon adjusting the duty cycle of PWM signal, which is consistent with our theoretical analysis in Section 3.2. Additionally, a wide range of dimming level is achieved in DCHO-OFDM. To expound, the proposed scheme is capable of achieving a fairly low or high dimming level, upon further increasing the value of the scaling factors, albeit at a cost to power inefficiency.

In Figure 6, we demonstrate the BER performance of DCHO-OFDM vs the variance of signal under different values of duty cycle, where a standard OFDM receiver is directly employed. Upon observing Figure 6, the attained BER of the proposed scheme first becomes better upon increasing the variance of signal, while it deteriorates afterwards due to the clipping noise generated by the limited dynamic range of LEDs. Additionally, we observe that the same BER performance is obtained with various values of duty cycle ρ, which indicates that the BER performance of DCHO-OFDM is invulnerable to the various duty cycles of PWM. This can be explained by noting that the dimming control in the proposed DCHO-OFDM scheme does not affect the information transmission, so that both the functionality of illuminations and communications in VLC can be achieved without disturbing each other. Hence, the symbols transmitted by the proposed DCHO-OFDM can be directly detected through a standard OFDM receiver without requiring additional operation of PWM signal detection.

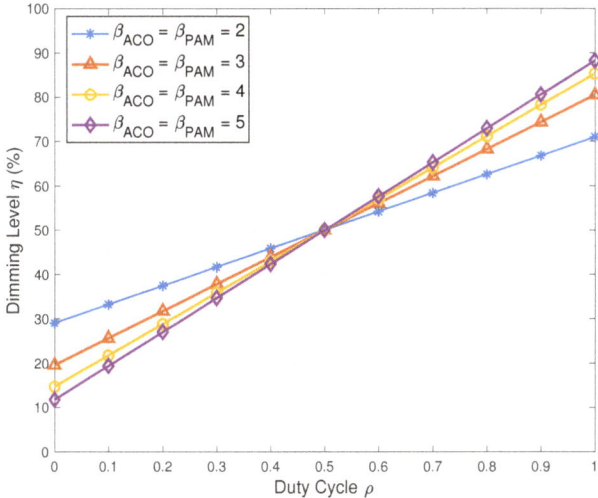

Figure 5. The performance of dimming level η vs duty cycle ρ, with various values of scaling factors.

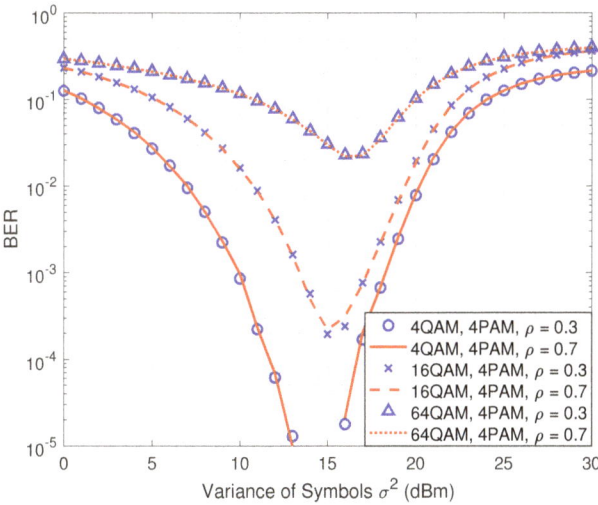

Figure 6. BER performance of the proposed DCHO-OFDM scheme under different variances of signal.

For a target BER of 2×10^{-3}, we further explore the spectrum efficiency of the proposed scheme, as depicted in Figures 7 and 8, in a scenario of high (-2 dBm) and of low (-8 dBm) noise power, respectively. For comparison, the spectrum efficiency of DCO-OFDM and the dimmable HACO-OFDM proposed in [22] are also provided. Upon observing, the achievable spectrum efficiency of DCHO-OFDM is superior to that of the conventional DCO-OFDM and HACO-OFDM in a wide range of dimming levels, where the transmission in DCO-OFDM is easily affected by dimming control due to the non-linear distortion. Furthermore, in contrast to HACO-OFDM scheme, while it has to detect ACO-OFDM signal first and then demodulatethe PAM-DMT signal, our proposed scheme can be detected upon utilizing the same receiver as the DCO-OFDM scheme, beneficially decreasing the complexity of receiver. Therefore, the proposed DCHO-OFDM scheme may be further cooperated with DCO-OFDM regime for a VLC system to satisfy the different transmission and illumination requirements, without increasing the complexity at the receiver side.

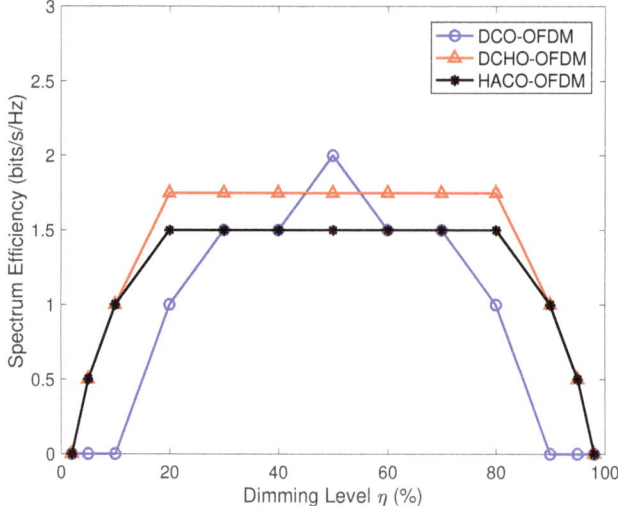

Figure 7. The achievable spectrum efficiency as a function of dimming level η, under a noise power of -2 dBm.

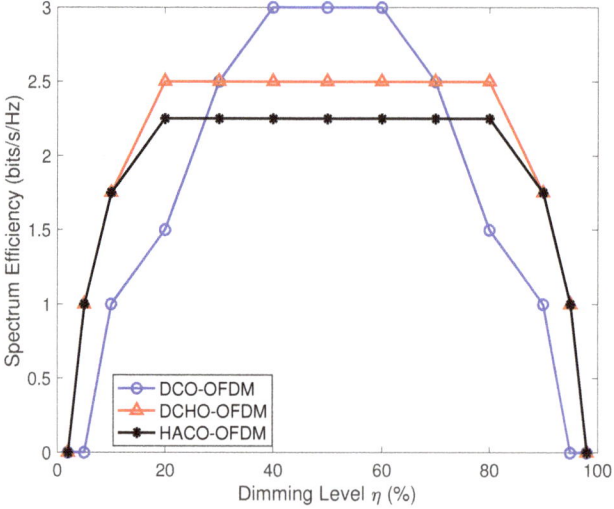

Figure 8. The achievable spectrum efficiency as a function of dimming level η, under a noise power of -8 dBm.

5. Conclusions

In this contribution, a novel DCHO-OFDM scheme was conceived for simultaneously achieving the dual functionality of data transmission and illumination in VLC. To achieve high spectrum efficiency, the bipolar unclipped ACO-OFDM signal was elaborately combined with the unipolar down-clipped PAM-DMT signal during the OFF state of PWM signal, while it was jointly transmitted with the upper-clipped PAM-DMT signal during the ON state. For the purpose of dimming control, a time-varying bias was applied to convert the superposed bipolar signals to unipolar ones, so that the PWM signal can be invoked without introducing interference. According to our simulations, a wide dimming range was obtained in DCHO-OFDM, with a high spectral efficiency and less fluctuation.

As the interference generated by the proposed scheme is orthogonal to the transmitted information, the demodulation of DCHO-OFDM can be easily executed upon employing a standard OFDM receiver, which may allow the proposed scheme to further cooperate with the DCO-OFDM scheme, fulfilling the dynamic communication and illumination requirements of the VLC system.

Author Contributions: Conceptualization, S.F.; Methodology, S.F.; Software, S.F.; Validation, H.F., Y.Z. and B.L.; Formal Analysis, S.F.; Investigation, H.F.; Resources, S.F.; Data Curation, S.F.; Writing–original Draft Preparation, S.F.; Writing–review & Editing, B.L.; Visualization, S.F.; Supervision, B.L.; Project Administration, H.F.; Funding Acquisition, B.L.

Funding: This research was supported in part by the National Natural Science Foundation of China under Grant 61571108, in part by the open research fund of National Mobile Communications Research Laboratory, Southeast University (No. 2019D18), in part by the Research Center of Optical Communications Engineering and Technology, Jiangsu Province (ZXF201902), and in part by the Fundamental Research Funds through the Central Universities under Grant JUSRP11919, in part by the Wuxi Science and Technology Development Fund (No. H20191001).

Conflicts of Interest: The authors declare no conflict of interest.

Appendix A

According to the definition of the upper-clipping, the upper-clipped PAM-DMT signal can be re-written as:

$$\lceil y_{PAM,n} \rceil_c = \frac{y_{PAM,n} - |y_{PAM,n}|}{2}. \tag{A1}$$

Upon taking the N-point FFT of signal $\lceil y_{PAM,n} \rceil_c$, we can get:

$$Y_k^c = \sum_{n=0}^{N-1} \lceil y_{PAM,n} \rceil_c \exp\left(-j\frac{2\pi nk}{N}\right)$$
$$= \underbrace{\frac{1}{2}\sum_{n=0}^{N-1} y_{PAM,n} \exp\left(-j\frac{2\pi nk}{N}\right)}_{W_1} - \underbrace{\frac{1}{2}\sum_{n=0}^{N-1} |y_{PAM,n}| \exp\left(-j\frac{2\pi nk}{N}\right)}_{W_2}. \tag{A2}$$

It can be found that the first term W_1 in (A2) is the FFT results of the unclipped signal $y_{PAM,n}$, which is the original transmitted PAM-DMT symbols carried by the imaginary parts of the even-indexed subcarriers. The second term of W_2 is viewed as the interference, imposed by the upper-clipping operation. Recall the property of $y_{PAM,n} = y_{PAM,n+N/2} = -y_{PAM,N/2-n} = -y_{N-n}$ [14], the second term of (A2) can be re-written as:

$$W_2 = \frac{1}{2}\sum_{n=1}^{N/4-1} \left[|y_{PAM,n}| \exp\left(-j\frac{2\pi nk}{N}\right) + |y_{PAM,N/2-n}| \exp\left(-j\frac{2\pi(N/2-n)k}{N}\right) \right.$$
$$\left. + |y_{PAM,N/2+n}| \exp\left(-j\frac{2\pi(N/2+n)k}{N}\right) + |y_{PAM,N-n}| \exp\left(-j\frac{2\pi(N-n)k}{N}\right) \right] \tag{A3}$$
$$= \frac{1}{2}\sum_{n=1}^{N/4-1} |y_{PAM,n}| \left\{ [1 + \exp(-j\pi k)] \left[\exp\left(j\frac{2\pi nk}{N}\right) + \exp\left(-j\frac{2\pi nk}{N}\right) \right] \right\}.$$

Since we have

$$\exp(-j\pi k) = \begin{cases} -1, & k \text{ is odd-valued} \\ 1, & k \text{ is even-valued} \end{cases}, \tag{A4}$$

the second term W_2 can be further expressed as:

$$W_2 = \begin{cases} 0, & k \text{ is odd-valued} \\ \sum_{n=1}^{N/4-1} |y_{PAM,n}| \cos\frac{2\pi nk}{N}, & k \text{ is even-valued} \end{cases}. \tag{A5}$$

Upon observing, it can be easily found that the introduced interference caused by the upper-clipping falls only on the real parts of the even-indexed subcarriers, where the odd-indexed subcarriers and the imaginary parts of the even-indexed subcarriers are free of contamination. As a benefit, the proposed DCHO-OFDM scheme is capable of remaining uncorrupted by the upper-clipped PAM-DMT signal.

Appendix B

Upon taking the N-point FFT of z_n, we obtain the frequency-domain sample Z_K for $k = 0, 1, \cdots, N-1$ as:

$$\begin{aligned} Z_k &= \sum_{n=0}^{N-1} z_n \exp\left(-j\frac{2\pi nk}{N}\right) \\ &= z_0 + z_{\frac{N}{4}} \exp\left(-j\frac{\pi}{2}k\right) + z_{\frac{N}{2}} \exp\left(-j\pi k\right) + z_{\frac{3N}{4}} \exp\left(-j\frac{3\pi}{2}k\right) \\ &+ \sum_{n=1}^{N/4-1} \left[z_n \exp\left(-j\frac{2\pi nk}{N}\right) + z_{\frac{N}{2}-n} \exp\left(-j\frac{2\pi(\frac{N}{2}-n)k}{N}\right) \right. \\ &\left. + z_{\frac{N}{2}+n} \exp\left(-j\frac{2\pi(\frac{N}{2}+n)k}{N}\right) + z_{N-n} \exp\left(-j\frac{2\pi(N-n)k}{N}\right) \right]. \end{aligned} \quad (A6)$$

According to Proposition 1, we have $z_n = z_{N/2+n}$ for $n = 0, N/4$ and have $z_n = z_{N/2-n} = z_{N/2+n} = z_{N-n}$ for $n = 1, \cdots, N/4-1$. Therefore, the Equation (A6) can be re-written as:

$$\begin{aligned} Z_k &= [1 + \exp(-j\pi k)] \left[z_0 + z_{N/4} \exp\left(-j\frac{\pi}{2}k\right) \right] \\ &+ \sum_{n=1}^{N/4-1} z_n [1 + \exp(-j\pi k)] \left[\exp\left(-j\frac{2\pi nk}{N}\right) + \exp\left(j\frac{2\pi nk}{N}\right) \right]. \end{aligned} \quad (A7)$$

If the value of k is odd, where $\exp(-j\pi k) = -1$, then we have:

$$Z_k = 0, \quad k = 1, 3, 5, \cdots, N-1. \quad (A8)$$

Otherwise if the value of k is even, where $\exp(-j\pi k) = 1$, the expression of Z_k is therefore given by:

$$Z_k = 2z_0 + 2z_{N/4} \exp(-j\frac{\pi}{2}k) + 2 \sum_{n=1}^{N/4-1} z_n \left[\exp\left(-j\frac{2\pi nk}{N}\right) + \exp\left(j\frac{2\pi nk}{N}\right) \right]. \quad (A9)$$

According to Euler formula, the real-part and the imaginary-part of Z_k are given by:

$$\Re[Z_k] = 2z_0 + 2z_{N/4} \cos\frac{\pi k}{2} + 4 \sum_{n=1}^{N/4-1} z_n \cos\frac{2\pi nk}{N}, \quad k = 0, 2, 4, \cdots, N-2, \quad (A10)$$

$$\Im[Z_k] = 0, \quad k = 0, 2, 4, \cdots, N-2. \quad (A11)$$

Upon observing Equations (A8), (A10) and (A11), it can be concluded that the interference generated by z_n is only imposed on the real part of the even-indexed subcarriers, leaving the odd-indexed subcarriers as well as the imaginary part of the even-indexed subcarriers uncontaminated.

References

1. Pathak, P.; Feng, X.; Hu, P.; Mohapatra, P. Visible light communication, networking and sensing: A survey, potential and challenges. *IEEE Commun. Surv. Tutor.* **2015**, *17*, 2047–2077. [CrossRef]
2. Bai, X.; Li, Q.; Tang, Y. A low-complexity resource allocation algorithm for indoor visible light communication ultra-dense networks. *Appl. Sci.* **2019**, *9*, 1391. [CrossRef]
3. Grobe, L.; Paraskevopoulos, A.; Hilt, J.; Schulz, D.; Lassak, F.; Hartlieb, F.; Kottke, C.; Jungnickel, V.; Langer, K. High-speed visible light communication systems. *IEEE Commun. Mag.* **2013**, *51*, 60–66. [CrossRef]
4. Rehman, S.; Ullah, S.; Chong, P.; Yongchareon, S.; Komosny, D. Visible light communications: A system perspective-Overview and challenges. *Sensors* **2019**, *19*, 1153. [CrossRef] [PubMed]
5. Feng, S.; Zhang, R.; Li, X.; Wang, Q.; Hanzo, L. Dynamic throughput maximization for the user-centric visible light downlink in the face of practical considerations. *IEEE Trans. Wirel. Commun.* **2018**, *17*, 5001–5015. [CrossRef]
6. Zuo, Y.; Zhang, J. A novel coding based dimming scheme with constant transmission efficiency in VLC systems. *Appl. Sci.* **2019**, *9*, 803. [CrossRef]
7. Feng, S.; Zhang, R.; Xu, W.; Hanzo, L. Multiple access design for ultra-dense VLC networks: Orthogonal vs non-orthogonal. *IEEE Trans. Commun.* **2019**, *67*, 2218–2232. [CrossRef]
8. Chen, Y.; Li, S.; Liu, H. Dynamic frequency reuse based on improved tabu search in multi-user visible light communication networks. *IEEE Access* **2019**, *7*, 35173–35183. [CrossRef]
9. Li, B.; Feng, S.; Xu, W.; Li, Z. Interference-free hybrid optical OFDM with low-complexity receiver for wireless optical communications. *IEEE Commun. Lett.* **2019**, *23*, 818–821. [CrossRef]
10. Chen, L.; Krongold, B.; Evans, J. Performance analysis for optical OFDM transmission in short-range IM/DD systems. *J. Lightw. Technol.* **2012**, *30*, 974–983. [CrossRef]
11. Dissanayake, S.; Armstrong, J. Comparison of ACO-OFDM, DCO-OFDM and ADO-OFDM in IM/DD systems. *J. Lightw. Technol.* **2013**, *31*, 1063–1072. [CrossRef]
12. Ranjha, B.; Kavehrad, M. Hybrid asymmetrically clipped OFDM-based IM/DD optical wireless system. *J. Opt. Commun. Netw.* **2014**, *6*, 387–396. [CrossRef]
13. Wang, Q.; Qian, C.; Guo, X.; Wang, Z.; Cunningham, D.; White, L. Layered ACO-OFDM for intensity-modulated direct-detection optical wireless transmission. *Opt. Express* **2015**, *23*, 12382–12393. [CrossRef] [PubMed]
14. Wang, T.; Hou, Y.; Ma, M. A novel receiver design for HACO-OFDM by time-domain clipping noise elimination. *IEEE Commun. Lett.* **2018**, *22*, 1862–1865. [CrossRef]
15. Sun, Y.; Yang, F.; Gao, J. Comparison of hybrid optical modulation schemes for visible light communication. *IEEE Photon. J.* **2017**, *9*. [CrossRef]
16. Feng, S.; Bai, T.; Hanzo, L. Joint power allocation for the multi-user NOMA-downlink in a power-line-fed VLC network. *IEEE Trans. Veh. Technol.* **2019**, *68*, 5185–5190. [CrossRef]
17. Gancarz, J.; Elgala, H.; Little, T. Impact of lighting requirements on VLC systems. *IEEE Commun. Mag.* **2013**, *51*, 34–41. [CrossRef]
18. Wang, Z.; Zhong, W.; Yu, C.; Chen, J.; Francois, C.; Chen, W. Performance of dimming control scheme in visible light communication system. *Opt. Exp.* **2012**, *20*, 18861–18868. [CrossRef]
19. Wang, Q.; Wang, Z.; Dai, L. Asymmetrical hybrid optical OFDM for visible light communications with dimming control. *IEEE Photon. Technol. Lett.* **2015**, *27*, 974–977. [CrossRef]
20. Stevens, N.; Strycker, L.D. Single Edge Position Modulation as a Dimming Technique for Visible Light Communications. *J. Lightw. Technol.* **2016**, *34*, 5554–5560. [CrossRef]
21. Hany, E.; Little, T. Reverse polarity optical-OFDM (RPO-OFDM): Dimming compatible OFDM for gigabit VLC links. *Opt. Exp.* **2013**, *21*, 24288–24299.
22. Yang, F.; Gao, J. Dimming control scheme with high power and spectrum efficiency for visible light communications. *IEEE Photon. J.* **2017**, *9*. [CrossRef]
23. Li, B.; Xu, W.; Feng, S.; Li, Z. Spectral-efficient reconstructed LACO-OFDM transmission for dimming compatible visible light communications. *IEEE Photon. J.* **2019**, *11*. [CrossRef]
24. Lee, K.; Park, H. Modulations for visible light communications with dimming control. *IEEE Photon. Technol. Lett.* **2011**, *23*, 1136–1138. [CrossRef]
25. Armstrong, J.; Lowery, A. Power efficient optical OFDM. *Electron. Lett.* **2006**, *42*, 1. [CrossRef]

26. Li, B.; Xu, W.; Zhang, H.; Zhao, C.; Hanzo, L. PAPR reduction for hybrid ACO-OFDM aided IM/DD optical wireless vehicular communications. *IEEE Trans. Veh. Technol.* **2017**, *66*, 9561–9566. [CrossRef]
27. Lee, S.; Randel, S.; Breyer, F.; Koonen, A. PAM-DMT for intensity-modulated and direct-detection optical communication systems. *IEEE Photon. Technol. Lett.* **2009**, *21*, 1749–1751. [CrossRef]
28. Islim, M.; Haas, H. Augmenting the spectral efficiency of enhanced PAM-DMT-based optical wireless communications. *Opt. Express* **2016**, *24*, 11932–11949. [CrossRef] [PubMed]

© 2019 by the authors. Licensee MDPI, Basel, Switzerland. This article is an open access article distributed under the terms and conditions of the Creative Commons Attribution (CC BY) license (http://creativecommons.org/licenses/by/4.0/).

Article

Shipborne Acquisition, Tracking, and Pointing Experimental Verifications towards Satellite-to-Sea Laser Communication

Dong He [1,2,3], Qiang Wang [1,2], Xiang Liu [1,2], Zhijun Song [1,2], Jianwei Zhou [1,2], Zhongke Wang [1,2], Chunyang Gao [1,2], Tong Zhang [1,2], Xiaoping Qi [1,2], Yi Tan [1,2], Ge Ren [1,2], Bo Qi [1,2], Jigang Ren [4,5], Yuan Cao [4,5] and Yongmei Huang [1,2,3,*]

1. Institute of Optics and Electronics, Chinese Academy of Sciences, No.1 Guangdian Road, Chengdu 610209, China; hedong@ioe.ac.cn (D.H.); wangqiang19750731@126.com (Q.W.); lx061990@163.com (X.L.); lala_xy@163.com (Z.S.); zjw96312@163.com (J.Z.); 15902875845@163.com (Z.W.); 13880872834@163.com (C.G.); 13880609238@163.com (T.Z.); 1qixiaoping@163.com (X.Q.); tandeman@126.com (Y.T.); renge@ioe.ac.cn (G.R.); qibo@ioe.ac.cn (B.Q.)
2. Key Laboratory of Optical Engineering, Chinese Academy of Sciences, Chengdu 610209, China
3. University of Chinese Academy of Sciences, Beijing 100049, China
4. Department of Modern Physics and Hefei National Laboratory for Physical Sciences at the Microscale, University of Science and Technology of China, Hefei 230026, China; jgren@ustc.edu.cn (J.R.); yuancao@ustc.edu.cn (Y.C.)
5. Chinese Academy of Sciences (CAS) Center for Excellence and Synergetic Innovation Center in Quantum Information and Quantum Physics, University of Science and Technology of China, Shanghai 201315, China
* Correspondence: huangym@ioe.ac.cn

Received: 1 August 2019; Accepted: 16 September 2019; Published: 19 September 2019

Abstract: Acquisition, tracking, and pointing (ATP) is a key technology in free space laser communication that has a characteristically high precision. In this paper, we report the acquisition and tracking of low-Earth-orbit satellites using shipborne ATP and verify the feasibility of establishing optical links between laser communication satellites and ships in the future. In particular, we developed a shipborne ATP system for satellite-to-sea applications in laser communications. We also designed an acquisition strategy for satellite-to-sea laser communication. In addition, a method was proposed for improving shipborne ATP pointing error. We tracked some stars at sea, achieving a pointing accuracy of less than 180μrad. We then acquired and tracked some low-Earth-orbit satellites at sea, achieving a tracking accuracy of about 20μrad. The results achieved in this work experimentally demonstrate the feasibility of ATP in satellite-to-sea laser communications.

Keywords: satellite-to-sea laser communication; acquisition, tracking and pointing; shipborne ATP; pointing error model

1. Introduction

Laser communication is a technology that uses a laser beam as a carrier to transmit information in space [1,2]. The laser beam used in space laser communication is emitted at an angle that is near the diffraction limit. The divergence angle of very narrow communication beams requires both communicating sides to maintain extremely high dynamic alignment accuracy. With the ultimate aim of realizing a global-scale laser communication for practical use, many significant achievements have been made in the past decades, such as the demonstration of links for building-to-building communication [3,4], airborne applications [5–7], inter-satellite applications [8,9], satellite-to-ground [10–16], and ship-to-ship communication [17–19]. However, currently, most of the experiments only involve space and ground communication links or short-distance communication between ships. Since water bodies occupy more than 70% of the earth's surface and given the rising demand for maritime communications, the global-scale communication network is bound to extend from the ground to the sea in the future. Huge volumes of data need to be transmitted by satellite. Thus, there is an urgent need to conduct experimental verifications of laser communication between satellites and ships at sea.

Satellite-to-sea laser communication mainly refers to data transmission between satellites and shipboards. Acquisition, tracking, and pointing (ATP) is the key technology used in laser communication, which requires the rapid acquisition and alignment of line-of-sight between two platforms and the establishment of communication links with high precision, high probability, and high dynamic tracking. The ATP system's rapid and high probability acquisition of a satellite terminal's line-of-sight under shipborne conditions has a direct impact on the communication time. Currently, optical links between ships have been verified. The link is usually initiated actively by an ATP at one end to cover an uncertain area with beacon light, then scanned or stared at by an ATP at another end to establish the link. This method of establishing an optical link is the most common method in laser communication, which requires high pointing accuracy. For short-distance communication links of tens of kilometers between ships or in air [20,21], the divergence angle and optical power of the beacon light can be improved, and the requirement of the ATP with respect to the pointing accuracy can be reduced to reduce the difficulty of acquisition. Another method for realizing short-range ship-to-ship laser communication is using modulating retro-reflectors (MRRs). An MRR link uses an active ATP at one end and a small semi-passive MRR at the other [22–25]. For one end of the link, MRRs no longer have tight pointing requirements. However, satellite-to-sea laser communication is very different from ground ATP systems. For ship-to-ship laser communications of tens of kilometers, the divergence angle of the uplink beacon is usually only a few milliradians due to resource constraints. The MMR method is less likely to achieve communication over hundreds or thousands of kilometers of links. Moreover, the complex nature of the ocean environment significantly increases the difficulty of acquisition and tracking by the ATP system. It is difficult for shipborne ATP systems to achieve high pointing accuracy to complete initial acquisition under conditions such as random and sharp waggling along with the sea, which poses a higher challenge to its technology. As a first step toward satellite-to-sea laser communication, it is essential to develop shipborne ATP and test its performance.

In this paper, we report the acquisition and tracking of low-Earth-orbit satellites using shipborne ATP and verify the establishment of optical links between laser communication satellites and a ship at sea. To improve the initial pointing accuracy of line-of-sight in the satellite-to-sea laser communication, we designed a calibration method for the shipborne ATP pointing error model. Some satellites were acquired and tracked by the ATP system, thereby demonstrating the acquisition and tracking performance of the ATP system at sea. A tracking accuracy of about 20μrad was obtained, together with a pointing accuracy of less than 180μrad. Our implementation provides a solid test for shipborne ATP, thus paving the way for a global-scale laser communication network involving space, ground, and sea.

2. Strategies for Establishing Optical Links

We designed an acquisition process for satellite-to-sea laser communication. The communication handshake is initiated by shipborne ATP. The terminals on each side of the communication drive their respective optical antennas to roughly align with each other's instantaneous positions according to the ephemeris and attitude information. Shipborne ATP has high-precision pointing ability and can transmit uplink beacon light to accurately cover the satellite terminal. After the satellite coarse Charge Coupled Device (CCD) detects the beacon light of the shipborne ATP, closed-loop tracking is carried out with high accuracy. The satellite terminal emits downlink beacon light. After the downlink beacon light is detected by the shipboard ATP coarse CCD, stable closed-loop tracking is conducted to complete the acquisition. After establishing a stable communication link between the satellite and the sea, laser communication is achieved until the satellite passes through. The acquisition strategy is summarized in Figure 1.

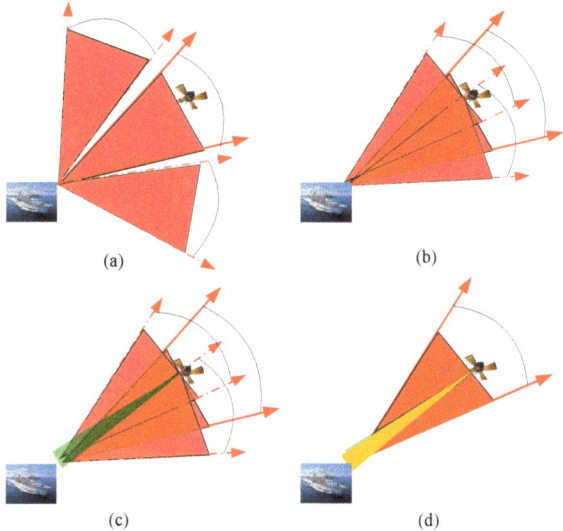

Figure 1. Schematic of acquisition and tracking. (**a**) The shipborne acquisition, tracking, and pointing (ATP) system initiates the acquisition process. The shipborne ATP and satellite activate their beacon lasers and point at each other using the predicted ephemeris. (**b**) The satellite detects the uplink beacon laser and initiates precise tracking. (**c**) The shipborne ATP detects the downlink beacon laser and initiates precise tracking. (**d**) Bidirectional tracking and locking between the satellite and shipborne ATP is established until the laser communication is terminated.

3. Method for Improving the Pointing Error

The acquisition protocol involves the shipborne ATP irradiating the satellite to initiate a link. The ATP is fixed on the ship. As the ship moves, the ATP is affected by sea waves, thus producing three periodic yaw, pitch, and roll disturbances, which cause the ATP's line-of-sight axis to wobble a few degrees. The divergence angle of the uplink beacon laser cannot be increased indefinitely; it is generally several milliradians and much smaller than the swaying amplitude. To achieve the acquisition of a satellite, the acquisition performance can only be improved by reducing the pointing error. The absolute pointing error is defined as the angular separation between the actual direction and the intended ATP line-of-sight [26]. In invisible satellite target acquisition, the pointing accuracy of the telescope is one of the important indices for the establishment of optical links. In our experiment, several factors contribute to the pointing error, including the attitude of the ship platform pre-compensation accuracy, installation error calibration accuracy, and systematic errors calibration accuracy.

3.1. Pre-compensation for Ship Platform Attitude

During the initial acquisition stage, the target coordinate value (A, E) in the Northeast celestial coordinate system is converted to (A_c, E_c) in the deck coordinate system. (A_c, E_c) is used to guide the ATP to open-loop pointing.

As shown in Figure 2a, the Northeast celestial coordinate system is defined as O-XYZ, where OX is due North, OY is due east, and OZ is due sky. The origin, O is the intersection of the azimuth axis and the pitching axis of the tracking equipment. The polar coordinates of the observation point, M are (A, E), while the projection of M on the OXY plane is N. A represents the azimuth angle, that is, the included angle of XON, which is positive when overlooking clockwise. E represents the elevation angle, which is the angle MON, where up is positive. L represents the distance from the tracking device to the observation point.

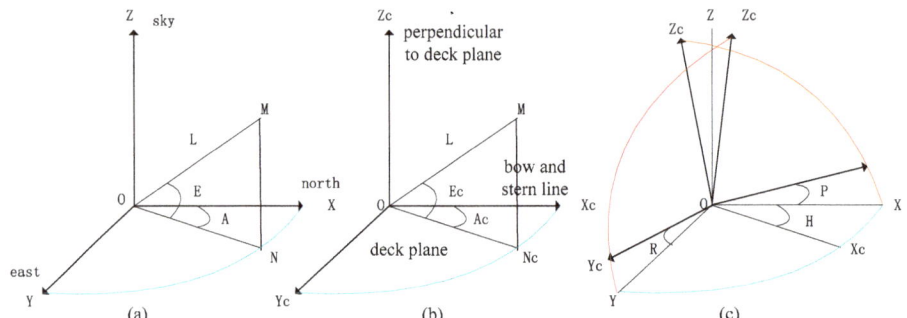

Figure 2. Schematic diagram of coordinate system and attitude angle. (**a**) The Northeast celestial coordinate system O-XYZ. (**b**) The deck coordinate system $O - X_c Y_c Z_c$. (**c**) Definition of attitude angle.

As shown in Figure 2b, the deck coordinate system $O - X_c Y_c Z_c$ is defined; the axis X_c is the fore-and-aft line, while the bow direction is forward. The axis OZ_c is positive clock wise; the axis Y_c is perpendicular to the fore-and-aft line in the deck plane, whereas the axis OZ_c is perpendicular to the deck plane, upward is positive. The origin O is consistent with the Northeast celestial coordinate system. The projection point of M on plane $OX_c Y_c$ is N_c. The polar coordinates of M in the deck coordinate system are defined as (A_c, E_c). A_c represents the azimuth angle of the deck, where clockwise is positive. E_c represents the elevation angle of the deck; upward is positive.

As shown in Figure 2c, the roll angle, R is the angle obtained by rotating the fore-and-aft line of the deck platform, and the port side rising is defined as a positive angle. The pitch angle, P is the angle included between the fore-and-aft line and the horizontal plane, and P is the positive angle when the bow is raised. The yaw angle, H is the rotation angle of the fore-and-aft line about the z-axis, and the clockwise angle is the positive angle.

When the shipborne ATP tracks a target, it is necessary to convert (A, E) of the target in the Northeast celestial coordinate system to (A_c, E_c) of the deck coordinate system for guidance and tracking.

In the Northeast celestial coordinate system, the transformation relationship exists between the polar coordinate system and rectangular coordinate system is given by Equation (1):

$$\begin{cases} x = L\cos E \cos A \\ y = L\cos E \sin A \\ z = L\sin E \end{cases} \quad (1)$$

In the deck coordinate system, the transformation relationship between the polar coordinate system and rectangular coordinate system is given by Equation (2):

$$\begin{cases} x_c = L\cos E_c \cos A_c \\ y_c = L\cos E_c \sin A_c \\ z_c = L\sin E_c \end{cases} \quad (2)$$

When the ship is affected by yaw, pitch, and roll at the same time, it shall be convertedin the order of yaw, pitch, and roll. The attitude rotation matrix, ξt is given by Equation (3):

$$\xi t = \begin{bmatrix} 1 & 0 & 0 \\ 0 & \cos R & \sin R \\ 0 & -\sin R & \cos R \end{bmatrix} \begin{bmatrix} \cos P & 0 & \sin P \\ 0 & 1 & 0 \\ -\sin P & 0 & \cos P \end{bmatrix} \begin{bmatrix} \cos H & \sin H & 0 \\ -\sin H & \cos H & 0 \\ 0 & 0 & 1 \end{bmatrix} \quad (3)$$

The conversion expression is shown in Equation (4) below:

$$\begin{bmatrix} x_c \\ y_c \\ z_c \end{bmatrix} = \xi t \begin{bmatrix} x \\ y \\ z \end{bmatrix}$$

$$= \begin{bmatrix} \cos H \cos P & \sin H \cos P & \sin P \\ -\sin H \cos R - \cos H \sin P \sin R & \cos H \cos R - \sin H \sin P \sin R & \cos P \sin R \\ -\cos H \sin P \cos R + \sin H \sin R & -\sin H \sin P \cos R - \cos H \sin R & \cos P \cos R \end{bmatrix} \begin{bmatrix} x \\ y \\ z \end{bmatrix} \quad (4)$$

It can be obtained the Equation (5) from Equations (1) and (2):

$$\begin{cases} A_c = \arctan(\frac{y_c}{x_c}) \\ E_c = \arcsin(\frac{z_c}{L}) \end{cases} \quad (5)$$

Thus, the target guidance value under the deck coordinate system can be obtained, as shown in Equation (6):

$$\begin{cases} A_c = \arctan\{\frac{\cos E\,[\cos R\,\sin(A-H) - \sin P\,\sin R\,\cos(A-H)] + \sin E\,\cos P\,\sin R}{\cos E\,\cos P\,\cos(A-H) + \sin E\,\sin P}\} \\ E_c = \arcsin\{\sin E\,\cos P\,\cos R - \cos E[\sin R\,\sin(A-H) + \sin P\,\cos R\,\cos(A-H)]\} \end{cases} \quad (6)$$

It can be seen from Equation (6) that the measurement accuracy for attitude (H, P, R) is one of the key factors affecting the accuracy of shipborne ATP pointing. The attitude error of the shipborne ATP is the main source of error that affects the orientation of the line-of-sight. We used an attitude pre-compensation unit (Figure 3), which used a combined inertial navigation equipment composed of a gyroscope, accelerometer, and GPS (GPS/INS) to measure the attitude of the ship. We made pre-compensation for the attitude by 200Hz, compensating for the wobble of the ATP line-of-sight axis caused by the swaying of the ship. The attitude error of the shipborne ATP is the main source of error that affects the orientation of the line-of-sight. The attitude of the ship is measured by the GPS/INS unit, which is converted into a compensation amount for the ATP line-of-sight axis for real-time attitude compensation using Equation (6).

Figure 3. Attitude pre-compensation unit.

The accuracy of the GPS/INS unit measurement is shown in Table 1. The angle measurement error for the GPS/INS unit roll and pitch direction root-mean-square (RMS) was $\delta_{Pitch} = \delta_{Roll} = 0.09$mrad (root-mean-square.) whereas the yaw direction angle measurement error was $\delta_{Yaw} = 0.18$mrad (RMS).

Table 1. GPS/INS unit attitude measurement accuracy.

Attitudeangle	Accuracy
Yaw	≤0.18mrad (RMS)
Roll and pitch	≤0.09mrad (RMS)

From Equation (6), the target guidance error caused by the GPS/INS unit can be obtained as shown in Equation (7), $\delta_{azimuth} = 0.18$mrad (RMS) and $\delta_{elevation} = 0.09$mrad (RMS).

$$\delta_{guidance} = \sqrt{\delta_{azimuth}^2 + \delta_{elevation}^2} = 0.202 \text{mrad} \qquad (7)$$

To improve the precision of the attitude compensation, the ship's attitude at the present moment as measured by the GPS/INS unit is used to predict the ship's attitude at a future time, while the guidance value of the target is pre-compensated. We used attitude velocity filtering and extrapolation to compensate the attitude lag. The running cycle of the software is 20 ms, and the attitude measurement data sent to the controller for execution needs three frames, a total of 60ms. Areal-time lag compensation of 60 ms was determined by testing the lag frame number on the swing table, so that the precision of the line-of-sight disturbance compensation reached the angular second level.

3.2. Installation Error Model

During the installation of the GPS/INS unit, inevitable installation errors are incurred between the measuring axis of the GPS/INS unit and the ATP pointing axis. If there is an error between the installation of the GPS/INS unit and the ATP, the attitude compensation accuracy will be affected and the pointing accuracy of the ATP will decrease. Therefore, after the installation of the GPS/INS unit and ATP, the rotation relationship between the platform coordinate system represented by the GPS/INS unit $O - X_p Y_p Z_p$ and the shipborne ATP deck coordinate system $O - X_c Y_c Z_c$ needs to be calibrated (Figure 4). When measuring the equipment attitude, it is necessary to convert the inertial attitude to the coordinate system of the equipment.

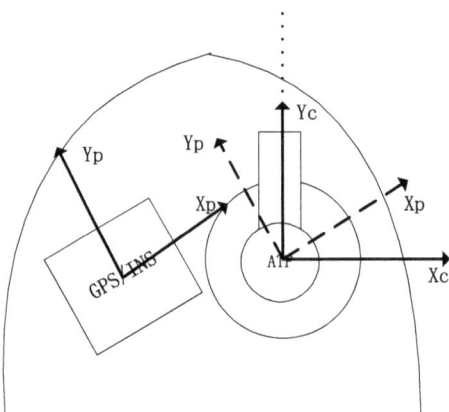

Figure 4. Error relation between the shipborne ATP and the GPS/INS unit.

The Equations (8) can be obtained from Equations (3) and (4), and Euler's theorem:

$$\begin{bmatrix} x_c \\ y_c \\ z_c \end{bmatrix} = \Psi t * \xi t * \begin{bmatrix} x \\ y \\ z \end{bmatrix} \qquad (8)$$

We can get from Equations (3) and (4), as shown in Equation (9).

$$\begin{bmatrix} L\cos E_c \cos A_c \\ L\cos E_c \sin A_c \\ L\sin E_c \end{bmatrix} = \Psi t * \xi t * \begin{bmatrix} L\cos E\cos A \\ L\cos E\sin A \\ L\sin E \end{bmatrix} \quad (9)$$

where (A_c, E_c) represents the target polar coordinate in the deck coordinate system, (A, E) represents the target coordinate value in the Northeast celestial coordinate system, ξt represents the attitude rotation matrix, and Ψt is the installation error matrix for the shipborne ATP and the GPS/INS unit.

Ψt is the parameter to be calibrated, Λ_c and Λ are coefficient matrices determined by experimental data, and the linear model of installation error can be expressed as Equation (10):

$$\Lambda_c = \Psi t * \xi t * \Lambda + \Delta \quad (10)$$

where, $\Lambda_c = \begin{bmatrix} L\cos E_c \cos A_c \\ L\cos E_c \sin A_c \\ L\sin E_c \end{bmatrix}$, $\Lambda = \begin{bmatrix} L\cos E\cos A \\ L\cos E\sin A \\ L\sin E \end{bmatrix}$, $\Delta = [\alpha, \beta, \theta]^T$.

The method of "star tracker" is used to evaluate the absolute pointing error of telescope, which is the main method to determine the absolute pointing error of ATP [27]. Because of the high precision of the known positions of stars in the sky, the identification of star signals provides a powerful tool for checking the absolute ATP alignment. In practical application, a star is considered a measurable point in the sky, and the least square fitting technology provides a solution for the absolute pointing error assessment [28].

In fact, we know the theoretical coordinates of the stars (A_i, E_i) in the Northeast celestial coordinate system. When measuring the star using the shipborne ATP, the measured value of the stars (A_{c_i}, E_{c_i}) can be obtained from the deck coordinate system. Several stars in the all-sky region were measured, and the error matrix, $\hat{\Psi} t$ was obtained by fitting with the least square method.

To sum up, the undetermined parameter $\hat{\Psi} t$ of the installation error model is calibrated by the least square method, and the residual error Δ is calculated by substituting into Equation (10). We regard residual error Δ as the systematic error of ATP. In the next section, the residual error is further corrected by establishing the shipborne system error correction model to achieve a high absolute pointing accuracy.

3.3. Systematic Errors Model

Due to the influence of machining, installation, and other factors, the telescope contains systematic errors. The error sources include the non-verticality of the horizontal and vertical axes, non-orthogonality of the visual and horizontal axes, and bending of the mirror tube and displacement of the optical axis. For ground telescopes, the research on absolute pointing accuracy correction method is relatively mature [29,30]. But at present there is no good method to complete the high precision systematic error calibration on the mobile platform.

We take the installation error model and systematic error model into consideration, and get the following expression, as shown in Equation (11):

$$\begin{cases} \Delta A_{c_i} = \hat{A}_{c_i} - A_{c_i} = f(A_{c_i}, E_{c_i}) + \varepsilon_i, i = 1, 2 \ldots, n \\ \Delta E_{c_i} = \hat{E}_{c_i} - E_{c_i} = g(A_{c_i}, E_{c_i}) + \tau_i, i = 1, 2 \ldots, n \end{cases} \quad (11)$$

where, $(\hat{A}_{c_i}, \hat{E}_{c_i})$ can be obtained from the equation $\hat{\Lambda} = \hat{\Psi} t * \xi t * \Lambda$, $(\Delta A_{c_i}, \Delta E_{c_i})$ are the observation error; $f(A_{c_i}, E_{c_i})$ and $g(A_{c_i}, E_{c_i})$ represent the approximate functions with the pointing angles (A_{c_i}, E_{c_i}) as unknowns, that is, the systematic error correction model; ε_i and τ_i are residual errors.

According to Figure 2, the zero position of the azimuth encoder is no longer aligned with due North, but with the zero position of the fore-and-aft line. Therefore, we consider the deck coordinate system as the reference when designing the error model.

The expression for the mount model [31] is shown in Equation (12):

$$\begin{cases} \Delta A_{c_i} = f(A_{c_i}, E_{c_i}) + \varepsilon_i = \sum_{j=1}^{m} d_j Y_j(A_{c_i}, E_{c_i}) + \varepsilon_i, i = 1, 2 \ldots, n \\ \Delta E_{c_i} = g(A_{c_i}, E_{c_i}) + \tau_i = \sum_{j=1}^{m} d_j \Gamma_j(A_{c_i}, E_{c_i}) + \tau_i, i = 1, 2 \ldots, n \end{cases} \quad (12)$$

where d_j, j = 1, 2, ..., m are the model coefficients for the m terms; $Y_j(A_{c_i}, E_{c_i})$ represents the function in azimuth residual of the azimuth, A_{c_i} and elevation E_{c_i} of star i, i = 1, 2, ..., n; $\Gamma_j(A_{c_i}, E_{c_i})$ represents the function in elevation residual of the azimuth, A_{c_i} and elevation E_{c_i} of star i, i = 1, 2, ..., n. The ATP mount model is presented in Table 2.

Table 2. The ATP mount model.

Term	Description	Azimuth Function (Y)	Elevation Function (Γ)
1.	Azimuth encoder offset	1	-
2.	Elevation encoder offset	-	1
3.	Azimuth axis tilt about fore-and-aft line	$-\cos A_c \tan E_c$	$\sin A_c$
4.	Azimuth axis tilt about axis Y_c	$-\sin A_c \tan E_c$	$-\cos A_c$
5.	optical axis misalign	$\sec E_c$	-
6.	Non-orthogonality of Azimuth and Elevation axes	$-\tan E_c$	-
7.	Azimuth bearing ellipticity (sin)	$\sin A_c$	-
8.	Azimuth bearing ellipticity (cos)	$\cos A_c$	-
9.	Azimuth bearing ellipticity (sin)	-	$\sin E_c$
10.	Azimuth bearing ellipticity (cos)	-	$\cos E_c$
11.	Telescope tube flexure	-	$\cot E_c$
12.	Azimuth encoder scale error	$A_c/2\pi$	-
13.	Elevation encoder scale error	-	$E_c/2\pi$
14.	Bi-periodic in azimuth	$\cos 2A_c$	-
15.	Elevation encoder stiction	-	$\sin A_c$
16.	Elevation bearing stiction	-	$E_c \sin A_c$

3.4. Installation and Systematic Error Model Calculation and Application Process

The installation and systematic error model can be obtained by the shipborne ATP's software automatically as shown in Figure 5a. ATP tracks and measures about 20 stars (magnitude less than 6, etc.) uniformly throughout the sky. Then the star's theoretical position (A_i, E_i), measurement position (A_{c_i}, E_{c_i}), and ship attitude (H_i, P_i, R_i) can be recorded. Once the measurements are available, the telescope will automatically track and measure the next star. The installation error matrix and systematic error model are solved by the least square method after all the stars measurement data are obtained. The whole process takes less than 10 min.

In the acquisition mode, the ideal point angle (A, E) in the Northeast celestial coordinate system is calculated according to the relationship between the target guidance position and the position of the ship. According to the measured attitude (H, P, R) and the installation error matrix Ψt, the guidance value (A_c, E_c) in the deck coordinate system can be calculated. Then, according to the pointing error correction model, the pointing error $(\Delta A_c, \Delta E_c)$ corresponding to the (A_c, E_c) pointing angle is solved. Finally, the pointing error $(\Delta A_c, \Delta E_c)$ is used to correct the actual pointing angle (A_c, E_c), and the correct pointing angle $(\hat{A}_c, \hat{E}_c)$ is obtained as shown in Figure 5b.

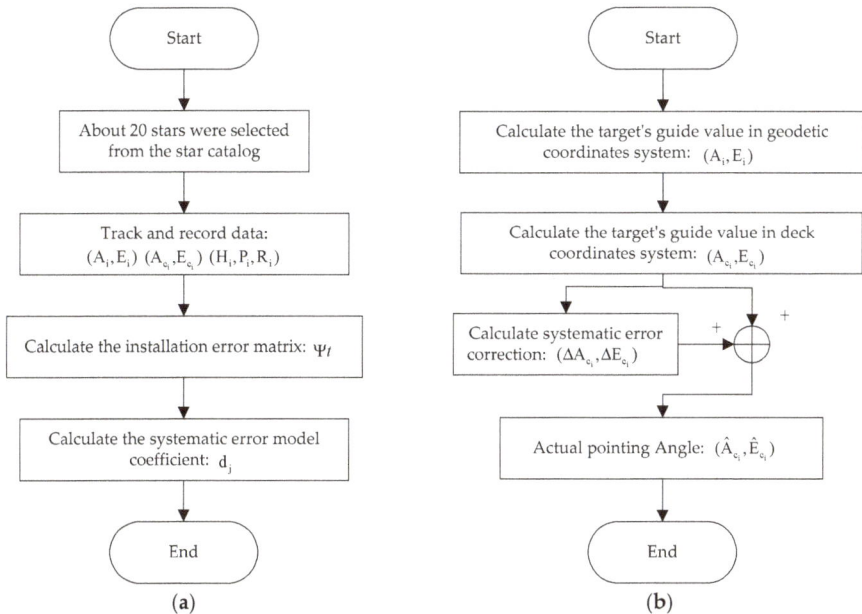

Figure 5. Installation and systematic error model calculation and application process. (**a**) Installation and system error model calculation flow chart. (**b**) Flow chart of actual use.

4. Experiment and Results

We developed a shipborne ATP system (Figure 6) mainly to verify the ability for acquisition and pointing, hence, a fast-steering mirror was not used at the fine pointing stage. The primary mirror in the ATP system is a reflecting Cassegrain telescope with an aperture of 75 mm and a focal length of 200 mm. The coarse-control loop included a two-axis gimbal telescope (azimuth rotation range of about −360° to +360°, elevation rotation range of about −5° to +90°) and a CCD camera (FOV of 0.97° × 0.91°, frame rate of 50 Hz).

Figure 6. The shipborne ATP system. The shipborne ATP system shows the uplink beacon, coarse camera, and the attitude pre-compensation unit.

We set up the ATP system terminal on an experimental ship at Dalian city. After the installation of the ATP system, the experimental ship sailed to an experimental site (38°48.027′ N, 121°50.037′ E) in the sea. We calibrated the ATP system's pointing ability by tracking and measuring a batch of stars one hour before the experiment. The residual error after installation error correction is shown in Figure 7 while the residual error after systematic error correction is shown in Figure 8. After obtaining the ATP system error model, a batch of stars were tracked and measured to verify the accuracy of the system error model. We used the shipborne ATP to track 15 stars with open-loop and obtained the pointing accuracy. The absolute pointing error obtained for the shipborne ATP azimuth was 117.8 μrad whereas the elevation was 128.1 μrad (Figure 9). The total pointing error of ATP is given by $\delta_{\text{pointing error}}$, as shown Equation (13).

$$\delta_{\text{pointing error}} = \sqrt{\delta_{\text{azimuth}}^2 + \delta_{\text{elevation}}^2} = 0.174 \text{ mrad} \tag{13}$$

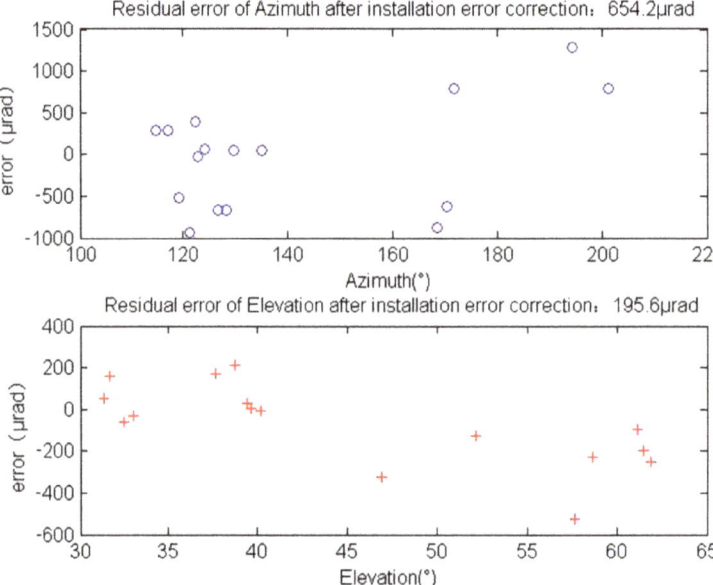

Figure 7. Residual error after installation error calibration. The azimuth residual error is 654.2 μrad while the elevation residual error is 195.6 μrad.

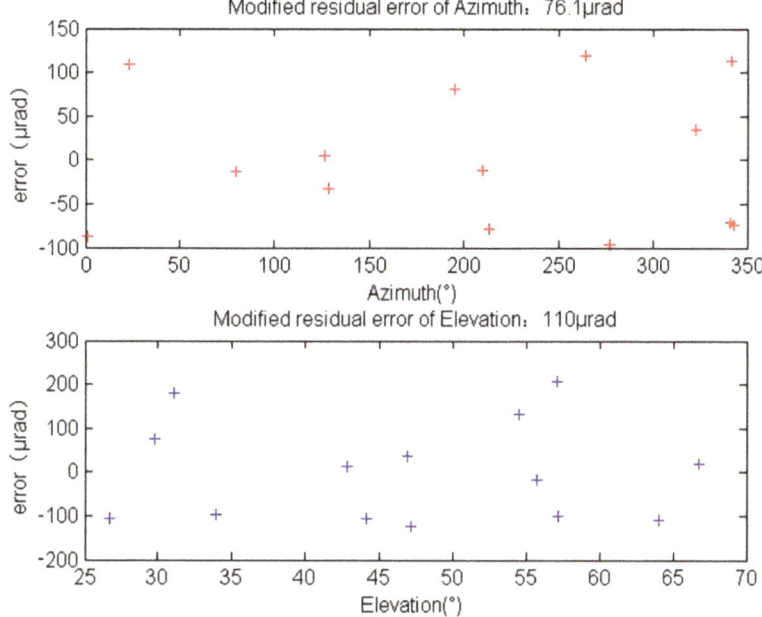

Figure 8. Residual error after systematic error correction. The azimuth residual error was obtained as 76.1μrad while the elevation residual error was 110μrad.

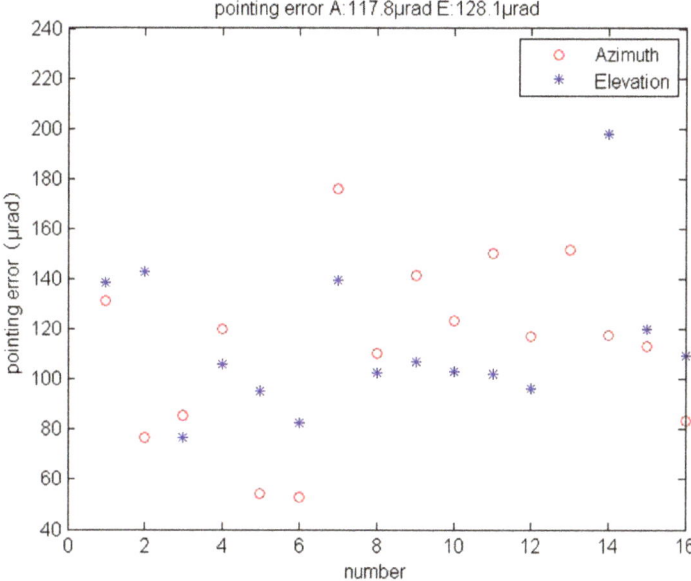

Figure 9. Pointing error of azimuth and elevation. Azimuth pointing error is 117.8μrad and elevation pointing error is 128.1μrad.

Figure 10 shows the performance of the shipborne ATP with respect to the acquisition and tracking of a satellite. The azimuth and elevation tracking errors were both less than 20μrad (RMS). The azimuth

115

initial acquisition error was 310μrad whereas the elevation initial acquisition error was160μrad. This error includes the orbital error of the satellite.

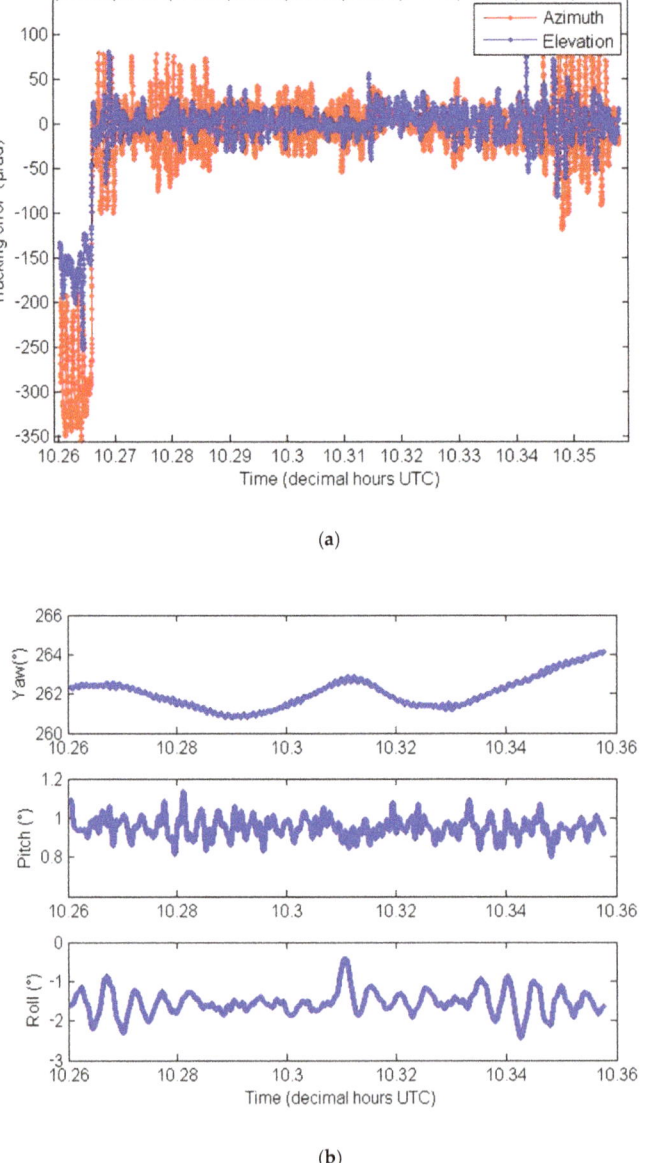

Figure 10. The shipborne ATP acquired and tracked the performance of a satellite. (**a**) The tracking error for the entire process and the initial error just appear in the field of view. The azimuth initial acquisition error was 310μrad whereas the elevation initial acquisition error was 160μrad. When the spot was in the tracking field, the azimuth tracking error was about 19.5μrad (RMS) while the elevation tracking error was about 14.6μrad (RMS). The tracking error in the stabilized time lasted from 10:15:58 to 10:21:28, about 330 seconds. (**b**) Ship attitude measurement. The yaw, pitch, and roll curves of the ship measured by the GPS/INS unit.

5. Conclusions

We tracked some stars at sea, achieving a pointing accuracy of less than 180μrad. We acquired and tracked some Low-Earth-orbit satellites at sea, achieving a tracking accuracy of about 20μrad. In this work, we have taken other sources of error into account, including position and alignment errors. The initial coarse orientation of the ATP was based on the satellite's predicted orbital position, with an uncertainty generally less than 200 m. A position error is less than 0.1mrad at 2000 km. An alignment error of less than 0.05mradcan be achieved between the beacon laser and tracking camera after adjustment. We can obtain a total initial open-loop pointing error of less than 0.324mrad, including position error (less than 0.1mrad at 2000 km), pointing error (less than 0.174mrad), and alignment error between the beacon laser and tracking camera (less than 0.05mrad). In accordance with two-dimensional Gaussian distributions, if the acquisition probability reaches 98.9%, the initial open-loop pointing error (RMS) of the shipborne ATP must be less than one-sixth of the beacon divergence angle. If the divergence angle of the ATP uplink beacon laser is greater than 2mrad, the shipborne ATP system open-loop pointing beacon laser spot can irradiate the satellite. Therefore, by verifying the pointing accuracy of the shipborne ATP and selecting the appropriate divergence angle for the uplink beacon, optical links between the shipborne ATP and the satellite can be established. We achieved significant experimental results for the shipborne ATP acquisition and tracking that will contribute to the realization of laser communication in free space between satellites and the sea.

Author Contributions: Conceptualization, D.H. and Y.H.; funding acquisition, Y.T., G.R., B.Q. and Y.H.; investigation, Q.W. and Z.S.; methodology, D.H., Q.W., Z.S., J.R. and Y.C.; project administration, Y.T.; resources, X.L., Z.S., J.Z., Z.W., T.Z., J.R. and Y.C.; software, D.H., Q.W., X.L., C.G., T.Z. and X.Q.; supervision, G.R., B.Q. and Y.H.; validation, X.L., J.Z. and C.G.; visualization, Z.W. and X.Q.; writing—original draft, D.H.; writing—review & editing, D.H.

Funding: This research was funded by the Natural National Science Foundation of China (NSFC), grant number (U1738204).

Acknowledgments: Thanks to our colleagues in the Key Laboratory of Optical Engineering, Chinese Academy of Sciences. Thanks for the financial support from Youth Innovation Promotion Association, Chinese Academy of Sciences.

Conflicts of Interest: The authors declare no conflict of interest.

References

1. Chan, V.W.S. Free-space optical communications. *J. Light. Technol.* **2007**, *24*, 4750–4762. [CrossRef]
2. Das, S.; Henniger, H.; Epple, B.; Moore, C.I.; Rabinovich, W.; Sova, R.; Young, D. Requirements and challenges for tactical free-space Lasercomm. In Proceedings of the MILCOM 2008-2008 IEEE Military Communications Conference, San Diego, CA, USA, 16–19 November 2008; IEEE: Piscataway, NJ, USA, 2009.
3. Kim, I.I. Wireless optical transmission of fast ethernet, FDDI, ATM, and ESCON protocol data using the Terra Link laser communication system. *Opt. Eng.* **1998**, *37*, 3143. [CrossRef]
4. Kim, I.I. Comparison of laser beam propagation at 785 nm and 1550 nm in fog and haze for optical wireless communications. *Proc. SPIE* **2001**, *4214*, 26–37.
5. Maynard, J.A.; Begley, D. Airborne laser communications: Past, present and future. *Proc. SPIE Int. Soc. Opt. Eng.* **2005**, *5892*. [CrossRef]
6. Fletcher, T.M.; Cunningham, J.; Baber, D.; Wickholm, D.; Goode, T.; Gaughan, B.; Burgan, S.; Deck, A.; Young, D.W.; Juarez, J.; et al. Observations of atmospheric effects for FALCON laser communication system flight test. *Proc. SPIE* **2011**, *8038*, 80380F-1–80380F-12.
7. Moll, F.; Horwath, J.; Shrestha, A.; Brechtelsbauer, M.; Fuchs, C.; Navajas, L.A.M.; Souto, A.M.L.; González, D.D. Demonstration of high-rate laser communications from a fast airborne platform. *IEEE J. Sel. Areas Commun.* **2015**, *33*, 1985–1995. [CrossRef]
8. Guelman, M.; Kogan, A.; Kazarian, A.; Livne, A.; Orenstein, M.; Michalik, H. Acquisition and pointing control for inter-satellite laser communications. *IEEE Trans. Aerosp. Electron. Syst.* **2004**, *40*, 1239–1248. [CrossRef]

9. Sodnik, Z.; Furch, B.; Lutz, H. Optical inter satellite communication. *IEEE J. Sel. Top. Quantum Electron.* **2010**, *16*, 1051–1057. [CrossRef]
10. Toyoshima, M.; Takayama, Y.; Takahashi, T.; Suzuki, K.; Kimura, S.; Takizawa, K.; Kuri, T.; Klaus, W.; Toyoda, M.; Kunimori, H.; et al. Ground-to-satellite laser communication experiments. *IEEE Aerosp. Electron. Syst. Mag.* **2008**, *23*, 10–18. [CrossRef]
11. Boroson, D.M.; Robinson, B.S. The lunar laser communication demonstration: NASA's first step toward very high data rate support of science and exploration missions. *Space Sci. Rev.* **2014**, *185*, 115–128. [CrossRef]
12. Robinson, B.S.; Boroson, D.M.; Burianek, D.A.; Murphy, D.V. The lunar laser communications demonstration. In Proceedings of the 2011 International Conference on Space Optical Systems and Applications (ICSOS), Santa Monica, CA, USA, 11–13 May 2011; IEEE: Piscataway, NJ, USA, 2011.
13. Wang, J.; Lv, J.; Zhao, G.; Wang, G. Free-space laser communication system with rapid acquisition based on astronomical telescopes. *Opt. Express* **2015**, *23*, 20655–20667. [CrossRef] [PubMed]
14. Wuchenich, D.M.R.; Mahrdt, C.; Sheard, B.S.; Francis, S.P.; Spero, R.E.; Miller, J.; Mow-Lowry, C.M.; Ward, R.L.; Klipstein, W.M.; Heinzel, G.; et al. Laser link acquisition demonstration for the GRACE Follow-On mission. *Opt. Express* **2014**, *22*, 11351–11366. [CrossRef] [PubMed]
15. Yin, J.; Cao, Y.; Li, Y.H.; Liao, S.; Zhang, L.; Ren, J.; Cai, W.; Liu, W.; Li, B.; Dai, H.; et al. Satellite-based entanglement distribution over 1200 kilometers. *Science* **2017**, *356*, 1140–1144. [CrossRef] [PubMed]
16. Liao, S.K.; Cai, W.Q.; Liu, W.Y.; Zhang, L.; Li, Y.; Ren, J.; Yin, J.; Shen, Q.; Cao, Y.; Li, Z.; et al. Satellite-to-ground quantum key distribution. *Nature* **2017**, *549*, 43–47. [CrossRef] [PubMed]
17. Rabinovich, W.S.; Moore, C.I.; Mahon, R.; Goetz, P.G.; Burris, H.R.; Ferraro, M.S.; Murphy, J.L.; Thomas, L.M.; Gilbreath, G.C.; Vilcheck, M.; et al. Free-space optical communications research and demonstrations at the US Naval Research Laboratory. *Appl. Opt.* **2015**, *54*, F189–F200. [CrossRef] [PubMed]
18. Gilbreath, G.C.; Rabinovich, W.S.; Moore, C.I.; Burris, H.R.; Mahon, R.; Grant, K.J.; Goetz, P.G.; Murphy, J.L.; Voelz, D.G.; Ricklin, J.C.; et al. Progress in laser propagation in a maritime environment at the Naval Research Laboratory. In Proceedings of the Free-Space Laser Communications V, SPIE Optics & Photonics 2005, San Diego, CA, USA, 31 July 2005; SPIE: Bellingham, WA, USA, 2005; Volume 5892, pp. 605–613.
19. Wu, R.; Zhao, X.; Liu, Y.; Song, Y. Initial pointing technology of line of sight and its experimental testing in dynamic laser communication system. *IEEE Photonics J.* **2019**, *11*, 1–8. [CrossRef]
20. Zhao, X.; Liu, Y.Q.; Song, Y. Line of sight pointing technology for laser communication system between aircrafts. *Opt. Eng.* **2017**, *56*, 126107. [CrossRef]
21. Klein, M.B.; Sipman, R.H. Large aperture Stark modulated retroreflector at 10.8 microns. *J. Appl. Phys.* **1980**, *51*, 6101–6104. [CrossRef]
22. Rabinovich, W.S.; Mahon, R.; Gilbreath, G.C.; Burris, R.; Goetz, P.G.; Moore, C.I.; Ferraro-Stell, M.; Witkowsky, J.L.; Swingen, L.; Oh, E.; et al. Free-space optical communication link at 1550 nm using multiple quantum well modulating retro-reflectors over a 1-kilometer range. In Proceedings of the Conference on Lasers and Electro-Optics, 2003, CLEO '03, Baltimore, MD, USA, 6 June 2003; IEEE: Piscataway, NJ, USA, 2004.
23. Rabinovich, W.S.; Goetz, P.G.; Mahon, R.; Swingen, L.A.; Murphy, J.L.; Ferraro, M.; Burris, H.R.; Moore, C.I.; Suite, M.R.; Gilbreath, G.C.; et al. 45-Mbit/s cat's-eye modulating retroreflectors. *Opt. Eng.* **2007**, *46*, 104001. [CrossRef]
24. Peter, G.G.; Mahon, R.; James, L.M.; Mike, S.F.; Michele, R.S.; Walter, R.S.; Ben, B.X.; Harris, R.B.; Christopher, I.M.; William, S.R.; et al. Modulating retro-reflector lasercom systems at the Naval Research Laboratory. In Proceedings of the Military Communications Conference, 2010-MILCOM, San Jose, CA, USA, 31 October–3 November 2010; IEEE: Piscataway, NJ, USA, 2010.
25. Majumdar, A.K. Modulating retroreflector-based free-space optical (FSO) communications. In *Advanced Free Space Optics (FSO)*; Springer: New York, NY, USA, 2015.
26. Mutilba, U.; Kortaberria, G.; Egaña, F.; Yagüe-Fabra, J.A. 3D Measurement Simulation and Relative Pointing Error Verification of the Telescope Mount Assembly Subsystem for the Large Synoptic Survey Telescope. *Sensors* **2018**, *18*, 3023. [CrossRef]
27. Huang, L.; Ma, W.; Huang, J. Modeling and calibration of pointing errors with alt-az telescope. *New Astron.* **2016**, *47*, 105–110. [CrossRef]
28. Tiziani, D.; Garczarczyk, M.; Oakes, L.; Schwanke, U.; van Eldik, C. A Pointing Solution for the Medium Size Telescopes for the Cherenkov Telescope Array. *AIP Conf. Proc.* **2017**. [CrossRef]

29. Cheng, J. *The Principles of Astronomical Telescope Design*; Springer: New York, NY, USA, 2009.
30. Donato, C.D.; Prouza, M.; Sanchez, F.; Santander, M.; Camin, D.; Garcia, B.; Grassi, V.; Grygar, J.; Hrabovský, M.; Řídký, J.; et al. Using stars to determine the absolute pointing of the fluorescence detector telescopes of the Pierre Auger Observatory. *Astropart. Phys.* **2007**, *28*, 216–231. [CrossRef]
31. Luck, J. Mount Model Stability. In Proceedings of the 14th International Workshop on Laser Ranging Instrumentation, San Fernando, Spain, 7–11 June 2004.

© 2019 by the authors. Licensee MDPI, Basel, Switzerland. This article is an open access article distributed under the terms and conditions of the Creative Commons Attribution (CC BY) license (http://creativecommons.org/licenses/by/4.0/).

MDPI
St. Alban-Anlage 66
4052 Basel
Switzerland
Tel. +41 61 683 77 34
Fax +41 61 302 89 18
www.mdpi.com

Applied Sciences Editorial Office
E-mail: applsci@mdpi.com
www.mdpi.com/journal/applsci

www.ingramcontent.com/pod-product-compliance
Lightning Source LLC
LaVergne TN
LVHW071958080526
838202LV00064B/6777